AF885748

STRATEGIC PLANNING FOR THE INDUSTRIAL ENGINEERING FUNCTION

STRATEGIC PLANNING FOR THE INDUSTRIAL ENGINEERING FUNCTION

Jack Byrd, Jr.
L. Ted Moore

VNR VAN NOSTRAND REINHOLD COMPANY
New York

Copyright © 1986 by Van Nostrand Reinhold Company Inc.

Library of Congress Catalog Card Number: 85-17977
ISBN: 0-442-26185-3

All rights reserved. No part of this work covered by the copyright hereon may be reproduced or used in any form or by any means—graphic, electronic, or mechanical, including photocopying, recording, taping, or information storage and retrieval systems—without permission of the publisher.

Manufactured in the United States of America

Published by Van Nostrand Reinhold Company Inc.
115 Fifth Avenue
New York, New York 10003

Van Nostrand Reinhold Company Limited
Molly Millars Lane
Wokingham, Berkshire RG11 2PY, England

Van Nostrand Reinhold
480 Latrobe Street
Melbourne, Victoria 3000, Australia

Macmillan of Canada
Division of Gage Publishing Limited
164 Commander Boulevard
Agincourt, Ontario M1S 3C7, Canada

15 14 13 12 11 10 9 8 7 6 5 4 3 2 1

Library of Congress Cataloging-in-Publication Data

Byrd, Jack
Strategic planning for the industrial engineering function.

Bibliography: p.
Includes index.
1. Industrial engineering. 2. Corporate planning.
I. Moore, L. Ted. II. Title.
T56.24.B97 1986 658.5′036 85-17977
ISBN 0-442-26185-3

PREFACE

This book is intended not only for industrial engineers but also for people who do the work of industrial engineers, managers who direct or work with I.E.'s, others who work with I.E.'s, and anyone interested in the I.E. philosophy of productivity.

Over years of working to help firms solve planning and operational problems the authors realized that the problems they were asked to address appeared to be the same problem, slightly revised but clearly identifiable. From that realization this book grew. The single problem the firms encounter is that they attempt to work in ways that are not operationally tenable.

Our consultations with companies have dealt with all levels of problems, from top management's strategic planning to plant management's shop floor control; yet the problems always have the same cause—actions are taken without rational justification. We have found companies with long lists of contradictory strategic objectives and companies that could not keep track of flows of materials; these examples represent the extremes of the operational levels of commonly encountered difficulties, but both result from the adoption of an irrational basis for operations.

Given the premise that irrationality is the common thread found in corporate problems, the natural question that comes to mind is HOW should rationality be brought to bear and WHO should do the job. No one individual can be expected to understand all details of any firm's activities, but someone with a clear vision of the factual, operational elements of productivity and profitability would be a prime candidate for the job. By training and experience industrial engineers fit the requirements for bringing bottom-up operational rationality to corporate planning and control.

However, an I.E. may lack insight into the broader perspectives motivating corporate actions that are generically referred to as strategic planning concepts. In presenting strategic planning concepts at national I.E. conferences we have encountered large audiences of interested listeners, suggesting a willingness on the part of the I.E. community to learn these perspectives.

We have also heard numerous upper level managers citing their wish for the involvement of assertive, strategically aware I.E.'s. These demonstrations of attitudes indicate that the time may be right for I.E.'s to learn broad corporate planning concepts. Combined with their knowledge of operations and their drive to improve productivity, I.E.'s could play a larger role in leading industry to greater efficiency and growth.

The topics in this book and their treatment are aimed at the audience cited above, people involved in corporate planning and management who are interested in both the work of I.E.'s and the logical selection of strategic plans. The thrust of topics has been influenced by our constant contact with managers, working I.E.'s, our students at West Virginia University, and alumni.

Our Industrial Engineering graduates are highly successful; their careers usually start with hands-on line responsibility and progress through staff level jobs to eventual positions in management. This sequence has been repeated so often that it seems the natural one. The individuals who follow this path undoubtedly acquire facility with the concepts presented herein during their progress, but this book may contribute to disseminating the ideas of industrial engineering as applied to strategic planning more broadly to other I.E.'s and interested readers.

This book does not attempt to cover the ground already covered many times by authors who have written strategic planning books stressing high-level management concepts and marketing theory. These are predominantly theoretical approaches to the topic of strategic planning and are of primary interest to specialists. We do not assume that I.E.'s will supplant specialists such as market analysts or computer programmers or manufacturing engineers and thus approach those disciplines only peripherally. However, the book is not highly technical in its treatment of I.E. concepts and should be readily accessible to anyone interested in rational strategic planning.

Two minor points about the book should be made. First, the discussion predominantly focuses on manufacturing operations rather than service firms. However, many of the topics addressed are readily applicable to the service sector—such topics as finance and facility decisions, for example. Secondly, there has been no attempt to employ the sexless pronouns that have come to have faddish usage; the masculine has been used as the common identifier for a person of either sex, thus avoiding the convoluted (and often erroneous) language that otherwise results. The authors are indebted to the numerous WVU I.E. alumni and students who have provided us with insights and information employed in the book. Similarly, members of the Department's Industry Advisory Organization have been especially helpful in keeping us up to date on a variety of managerial issues. We would also like to thank Van Nostrand Reinhold's staff for their technical assistance in preparing this book on computer disks.

CONTENTS

PREFACE v
INTRODUCTION ix

PART ONE: LEVELS OF STRATEGY IN CORPORATIONS 1

1. Long-term Planning 3
2. Short-term Planning 19
3. The Failures of Strategic Planning 33

PART TWO: THE POTENTIAL ROLE OF INDUSTRIAL ENGINEERS IN LONG-TERM PLANNING 41

4. Facilities Decisions 43
5. Focus of Direction 53
6. Acquisition and Expansion 67
7. Research and Development 82
8. Finance 97
9. Supply Sources 111
10. Distribution 121
11. Marketing and Approach to Competition 131
12. Organizational Structure 141

PART THREE: THE POTENTIAL ROLE OF INDUSTRIAL ENGINEERS IN SHORT-TERM PLANNING 153

13. Product Positioning 155
14. Product Development 163
15. Product Mix 172

16. Methods of Production 183
17. Integrated Planning and Control Systems 193
18. Distribution Procedures 202
19. Inventory Planning and Control 213
20. Scheduling 222
21. Facilities Utilization 234
22. Marketing Procedures 245

PART FOUR: ALLOCATION OF INDUSTRIAL ENGINEERING RESOURCES 253

23. A New Charter for Industrial Engineers 255

BIBLIOGRAPHY 263

INDEX 267

INTRODUCTION

Strategic planning is such a broad topic that it naturally means different things to different people. The term comes from military terminology in which a campaign with an overall goal of victory requires large-scale concepts as a guide toward that victory. These large concepts are called *strategy*; the specific procedures used to implement strategy are called *tactics*. In chess a strategy might be to develop pieces rapidly; pawn moves in the middle of the board might be the requisite tactics. In the Pacific campaign in World War II, the Allied strategy was to island-hop all the way to Japan. The Allies adopted amphibious beach assaults, long-range bombardments, air attacks, and naval blockades as tactics.

In the business world the policies that dictate competitive strategy are determined to some degree by the health of the company. A struggling company may be solely interested in survival; healthy companies usually seek higher profits. The past two decades in American business have been a virtual laboratory demonstration showing the results of experiments represented by diverse strategies spawned by these two corporate styles.

The early 80's have featured the activities of struggling companies. Their struggles may not be against competitive enemies; they may simply be suffering a lack of demand for their goods or services. During the recession of the 70's and 80's consumers were reluctant to buy many items they would have bought previously and their hesitancy weakened numerous firms.

Faced with this situation, the usual corporate strategy is to cut costs and lower the company's breakeven point. However, cutting production costs per unit is difficult in a period of low demand, so the commonly employed cost-reduction tactics are the reductions of workforce and inventories. Automotive firms have adopted this course, reducing inflated staff sizes to reduce fixed costs.

One ordinarily thinks of strategic planning as the mapping of plans for overcoming corporate competitors—plans typically encountered in a growing company. Strategies for increasing profits usually consist of making existing products more profitable or of moving into more profitable product lines.

The two fundamental strategies cited above can be summarized as cost cutting and profit enhancement, and these have been the basic strategies for centuries, but there is a difference between the way strategy was chosen long ago and the way it is chosen now. The difference has been caused by two primary factors: the power of new computational and data manipulation tools and the rapidity with which both products and demands now change. For example, between 1973 and 1982 the number of new food and drug products introduced into the market *per year* almost doubled and exceeded 1500 products in 1982.

The modern approach to strategic planning uses tools and ideas nonexistent for practical purposes 30 years ago. Many mathematical theories, useful in devising corporate plans, have only become of practical value within the past 30 years with the advent of inexpensive computation. Statistical forecasting and estimation procedures, linear programming, and simulation modeling are a few of the approaches that I.E.'s commonly employ in their planning work that have become readily implementable in formulating strategic plans. The advent of powerful microcomputers now permits personnel at all levels of companies to experiment with planning scenarios by use of spreadsheet software. This new environment is making managerial personnel more receptive to the advantages that computer tools offer the planning activity. I.E.'s have long been familiar with the newly accepted mathematical procedures, and this familiarity is one strong point they can bring to strategic planning.

The growth of computer use has changed the way most things are done throughout large corporations and has affected smaller firms to a growing degree. Not only has the planning process been affected; so have operating methods. Computers now permit on-line control of systems, numerical control of machines, scheduling and inventory control by use of integrated manufacturing systems, and flexible computer design of products and methods. In short, all operating systems that fall in the categories that rely heavily on I.E. expertise have been sharply affected by the computer revolution in recent years, and this fact points to the need for I.E. involvement in planning.

"New" ideas, or at least ideas that are now in vogue but have not always been, have also changed how companies are run and how plans are laid. Just-in-time production and experience curve theory are two ideas that appear to be new and that have attracted considerable attention in the MBA curricula and the business magazines. And receiving even greater attention has been the "new" idea of strategic planning.

Another recent change is management's growing consciousness of the need for formal strategies. This may be due to changing education in business schools, or it may be a natural result of competitive forces. In any event, strategic planning has become a distinct discipline and has spawned strategic planning consultants, new planning categories in corporate organization charts, and software specifically written for strategic planning applications.

During the 1970's consulting firms were founded solely to help companies plan strategy, e.g., the Boston Consulting Group [24]. It is becoming more common for larger corporations to perform their own strategic analyses internally, but consulting firms appear to have served to educate management in the concepts and methods of strategic planning. A manager without a grand strategy must now feel inadequately prepared.

The level of management at which strategic planning has characteristically taken place is the highest, for example GM's famous 14th floor. Participation in this approach to corporate strategy has usually been limited to upper management and staff, virtually excluding line managers. However, there has been a recent change in management's thinking about strategic planning; the current stress is on implementation of plans. As a result the levels of personnel and the levels of decision making attendant to strategic planning are changing.

The changes are sufficiently broad to have been featured in an article in *Business Week* [61]. The article cited numerous instances of companies turning away from the employment of strategic planning consultants and corporate strategic planning officers and adopting shorter lines of decision making. This revision in approach is summarized in the article as a new emphasis on "implementation." The immediate consequence of changing to an implementation-based mode of strategic planning is that people directly responsible for acting as the plans dictate will be expected to contribute to their development; in turn they will be held acountable for failures and will gain the satisfaction and rewards of their successes.

The new emphasis on the closer involvement of day-to-day line managers in casting strategy also caught the notice of the *Washington Post* [63]. The *Post* article dealt primarily with General Electric, a company long known for its enthusiasm for formal planning, although it too is now adopting a policy of shortening lines of control—moving decisions closer to the shop floor.

In both articles the motivating factors seem to be the need for timely decisions and the need for plans to be truly implementable. Both needs are best satisfied by people who fully understand the operating environment. The change in management's orientation to the strategic planning process is leading companies to educate line managers to the concepts and policies that dictate their strategic plans, hoping that this new knowledge, coupled with the line managers' skills will make them into effective participants in the planning/implementation process.

These changes present I.E.'s with excellent opportunities to participate in strategic planning roles for which they are uniquely equipped. The reasons for the changes in managerial thinking about strategic planning and the resulting opportunity for I.E.'s are found in basic economic and operational principles—how much things cost and how the operations work. These are the two dominant topics of I.E. expertise. Consider the role of costs.

Strategic planning of cost improvement and profit enhancement undertaken by top management requires that a few assumptions be made about what will happen once the developed strategies are implemented. Basic assumptions are not often articulated because they seem natural. They include the idea that market forces are going to work about the way they have previously and the idea that costs will evolve in a readily predictable fashion.

Costs constitute the dominant element of strategic planning because cost reduction is the dominant concern of a cost strategy, and a profit enhancement strategy requires that costs per unit either remain the same or decline. Since costs are of great concern, theoretical analyses of them have received considerable notice during the past decade. One such theoretical topic is that of the experience curve.

In many closely scrutinized industries it has been found that product costs per unit decline as more and more of the product is manufactured, although due to the effects of inflation many industries must have their current costs deflated and expressed in constant dollars before this result can be observed. This declining cost per unit, graphed against cumulative production, produces the *experience curve*. (See Figure 1.) Its lesson is that a combination of economic forces, learning, and technological substitution tend to decrease constant dollar costs as production continues.

Economic theory readily explains the industrywide experience curve. As competition for market share for a given product becomes strong, cost becomes the determinant of sales because prices cannot be reduced below costs for protracted periods. Thus, the producer that can reduce its cost per unit the fastest

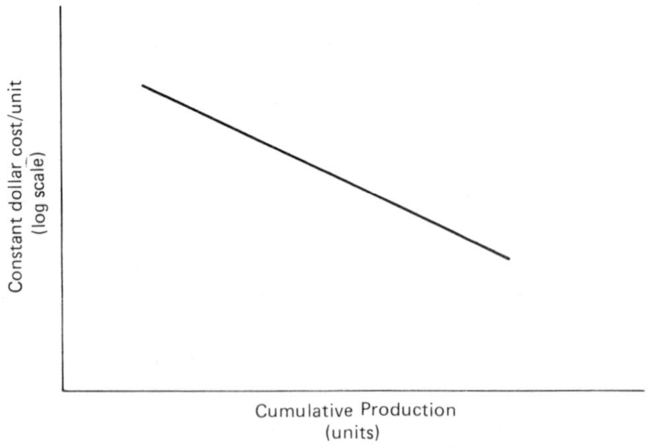

Figure 1. Experience Curve.

has the edge on competitors; it can sell at the lowest price and still make a profit. The higher cost producers will be forced to leave the market as will higher cost technologies.

The suggestion that inefficient producers and technologies would be crowded out of markets is not a novel idea. This tendency is a result of the diversity among producers. If there were only one way to make a product and only one possible cost per unit, market forces would encourage the existence of a large number of local suppliers.

What is revelatory is that there is sufficient diversity of available methods, materials, and applications of labor within a single firm to permit a reduction in costs per unit of product much like that experienced by an entire industry. The implication is that, by careful thought and monitoring, a company can plan a strategy of internal cost reduction on its major products. One might argue that the experience curve concept appears to be no more than a combination of ideas about productivity improvement and economies of scale, but the novel aspect of experience curve theory is that a company can expect a predictable cost decline as cumulative production mounts.

Several other recent stimuli have made costs more central to planning. One stimulus is the growing influence of managerial accounting as a formal discipline. As accountants are trained more in the practical procedures of producing cost numbers that managers need as well as numbers required by the public accounting function, better cost decisions can be made.

A second stimulus is the recent improvement of techniques for studying production costs at the design stage of building a new production facility. Especially powerful is the technique of *life cycle costing* (LCC), which explicitly models the cost tradeoffs inherent in a plant design.

The key requirement of successful cost-oriented strategies is that the procedures for planning and monitoring of costs must be developed and maintained by personnel who are equipped with the breadth of background that will allow them to understand the operational details, the mathematical and computational aspects of both operations and strategic planning, and the broad objectives of the work. Most upper managers, those people most likely to make strategic decisions, are often not equipped with the requisite breadth of training and background. Neither is the academically trained management scientist, because he is unlikely to have the necessary knowledge of operations. The best candidate for an individual able to fully partake in planning, analysis, implementation, and monitoring is the Industrial Engineer, but if he is to fulfill his potential in this role he must educate himself to the concepts and applications of strategic planning.

The failure of strategic planning as it is frequently practiced derives from the absence of involvement by those who will ultimately be responsible for implementing key elements of the plan. This fact is a major reason companies are trying to involve line personnel more in the total strategic planning process. Disappearing from the scene will be the ''ivory tower'' management scientist, the high-paid consultant unfamiliar with the business he helps plan, and the strategic planning manager who has never served a line function. The way appears to be opening for aggresive I.E.'s to assert themselves into a rational new version of strategic planning; they have an excellent opportunity to develop themselves into the kind of personnel, possessed of a broad perspective, who can understand the broad issues that implementation-based planning will coordinate.

Modern industrial management has suffered from overspecialization, at least in the U.S. Within the hierarchical structure of an industrial organization, managers tend to adopt specialization of level, that is, a manager at a particular level in the hierarchy learns the tasks and skills peculiar to that one level. If he advances to a higher level, he learns that level's job, etc. The detrimental aspect of this situation is that most upper managers start in staff-level jobs and never learn the jobs below their entry level.

Companies in Japan [22] are said to develop managers who do not have the limitations of overspecialization. The Japanese method for developing broad knowledge in managers is to move people from one specialty to another until they have learned each major function. A person being groomed in this manner may spend some years in each of the Accounting, Finance, Marketing, Personnel, and Production divisions. Few American companies have had success with this approach to training, largely because American workers are more likely to change jobs to obtain a promotion than are Japanese workers. The more common American method of developing managers is to promote line personnel up through the line and then the staff ranks. A person who rises to a high level in this fashion at least understands all the jobs below him. There is one particular individual who can be expected to have started his professional life in line management at the operational level—the Industrial Engineer.

Management theorists explain the danger presented by managers who do not understand operational functions by saying that these managers become symbol manipulators, issuing directives about operations that have no more than symbolic significance to them. The natural result is that many directives made in ignorance require operational activities that are inadvised, impractical, or impossible. What is needed to resolve these difficulties is an interlocutor between the operational and strategic levels of corporate decision making who understands both levels, and the most likely candidate for this role is the Industrial Engineer.

The reasons I.E.'s fit this role so well derive from their training, career paths, and their activities within the corporation. The typical I.E. curriculum teaches about manufacturing processes; cost concepts; planning, modeling, and control of operations; and wages and incentives. Many I.E.'s take summer jobs during their school years in a manufacturing environments similar to that in which they would like to work. A typical graduate I.E.'s first job is on the plant floor. His path of advancement will take him through a plant manager job, and then he will enter corporate management. Many I.E.'s on this kind of career path eventually become chief officer of the company. This sequence of training and experience makes the I.E. an excellent candidate to fill the role of interlocutor between strategic and operational aspects of his company.

To seize this position in the organization an industrial engineer will probably have to take the initiative. Not only will he need his formal training and his line experience, he will have to learn the concepts and procedures of strategic planning. Only then will he be ready to take the role that seems so naturally to require him. This book is intended to provide information that will help prepare an industrial engineer for his modern function.

PART ONE

LEVELS OF STRATEGY IN CORPORATIONS

Corporate strategy can be divided into two categories: strategy for the long-term and strategy for the short-term. This division is necessary to our discussion if the major operational aspects that affect strategic planning are to be fully discussed. Since strategic needs of individual companies vary greatly, it is difficult to provide a noncontroversial listing of subclassifications; a serious concern at one company might be irrelevant at another. However, pointed discussions of the specific types of concerns that affect corporate strategies are facilitated by breaking strategic planning into subclassifications. All specified in this book are important to many companies. Before examining each identified aspect of strategic planning, let us review the long-term nature of strategy selection.

1
LONG-TERM PLANNING

We must comment at this point on the possible confusion between the concepts of strategic and long-term planning. Long-term planning exists in virtually all firms regardless of whether there is any strategic planning. Long-term decisions must be made, especially about sources of capital, in every operating environment, but these decisions are not necessarily founded on articulated strategy. On the other hand, most strategic plans have important long-term implications, so it is easy to see how the two terms might be thought to be interchangeable.

This book is about strategic planning, so the only variety of long-term planning considered herein is that based on strategic decisions. However, strategy also requires that people and facilities be able to make rapid changes in tactics, so strategies must be viewed as also containing short-term elements. In this book references to short-term planning will mean those aspects of strategic planning that are dictated by foreseeable short-term needs.

FACILITIES DECISIONS

Facilities decisions are almost always capital intensive, which makes these decisions inherently long-term. Few capital investments can be viewed as recoverable over periods shorter than several years. In addition, once a significant facility investment has been made, the fixed nature of this part of the company's capital base implicitly focuses part of the company's activities toward the capabilities of that portion of its facilities.

A good example of the inertia that facility decisions build into a firm is provided by the U.S. steel industry. Once a steel mill is erected, it is not adaptable to ends other than the production of steel products. The natural response of management in the steel industry to declining profitability of steel mills has been diversification, but the influence of the large fixed investment in traditional steelmaking facilities led most steel companies to add on to the old mills rather than to concentrate on the efficient production of specific product lines at new special-purpose mills. The result has been larger and larger steel mills that turn out an increasingly complex diversity of specialized products. Rails, strip, plated products, special alloys, tinned products, and other steel products for special needs have been added at fixed sites. This kind of product proliferation at single sites is not usually a wise policy, but for steel it was an almost inevitable strategy due to the influence of prior facility investments. Thus, while the old traditional mills ran into serious problems, special-purpose "mini-mills" continued to make a profit even in the face of severe economic conditions.

One reason for the lack of steel industry interest in modernizing its facilities is the "product/process" dichotomy [70]. Strategic plans in many industries have often been based on marketing (product) objectives to the exclusion of production (process) considerations. This division of viewpoints can be seen in the early influence on strategic planning of marketing theorists such as Levitt [18]. The marketing view stresses product innovation and market development. Applied to the steel industry, this view would suggest the development of a wider range of products at the integrated steel mills. The implied proliferation appears to have taken place.

The Japanese in recent years have taken the process-oriented view, which is commonly held by people trained in understanding production operations, people such as industrial engineers. Japanese steel mills make high-quality product at lower cost, largely because the Japanese mills employ more efficient, modern technology, especially continuous casters. Once the American steelmakers had understood why their competitive position had declined, their viable responses were cost cutting and replacement of outmoded equipment and procedures.

Unfortunately for the U.S. manufacturers, their ability to finance capitalization of the magnitude needed to recover their preeminence was absent. Cost cutting has been accomplished by closing unprofitable plants, eliminating product lines, and selling assets. As a result the American steel industry has shrunk to less than two-thirds of its former capacity. The plight of American steel appears to be rooted in overemphasis of product strategy and lack of attention to process concepts. The lesson in this story is that process-conscious personnel, such as I.E.'s, should be involved in strategic planning in addition to product-oriented personnel.

The consequences of facility decisions are not always so dire, but they may influence decisions far into the future, which makes them extremely important. Most facility decisions, if handled properly in the context of long-term company policy, can offer opportunities that no other type of decision can. Consider the following instance.

Some companies that require large amounts of fuel for heating (usually the heating of boilers to raise process steam or primary heating as in kilns) have made facility decisions that have improved their ability to respond to changing fuel market conditions. These companies have acquired the ability to use alternative fuels. With minor adjustments they can switch from oil to coal or from oil to gas, depending on which fuel is more economical at the time. Once this type of investment has become recognized as a sunk cost, it provides cost-free flexibility for many years. All decision makers needed was the willingness to invest in the proper kinds of boilers, burners, and storage sites for oil and coal and sufficient foresight to line up flexible fuel contracts.

The most common facility decisions are those dealing with replacement or retirement of old sites or equipment, revision of capacities at one or more sites, and the purchase or construction of entirely new facilities. All these decisions can affect the company's capacity, diversity, flexibility, profitability, and overall ability to compete.

No decisions are made in a vacuum. Other categories of decision making will affect facility decisions, and that is one reason planning categories must be viewed as arbitrary. For instance, a decision to enter a specific market or to change the focus of the business can implicitly dictate a facility plan.

FOCUS OF DIRECTION

Possibly the most important influence on the selection of strategies in any manufacturing company is its overall philosophy. All its important decisions spring from its philosophy, which may be explicit or implicit. The philosophy dictates the firm's focus of direction. A summary of the beliefs that guide a company may be long or short. Some companies are fond of saying that their objective is to serve the public. Other firms maintain a lengthy list of rules and criteria. A manufacturing firm's philosophy establishes which business it believes itself to be in, which markets it is attempting to serve, how it approaches product development, etc. Creating, maintaining, and changing this philosophy is the primary job of the chief executive officer (C.E.O.). An effective philosophy is needed to provide the firm with a well-conceived focus of direction.

The problem of poor focus in manufacturing firms was described forcefully by Skinner [69], who also explained the source of the problem, a lack of holistic viewpoints being brought to the planning process. Here is a strong motivation for the I.E. to educate himself to the principles and concepts of strategic planning; coupled with his ready comprehension of operations and costs, these concepts should provide him with a firm basis for holistic judgment. He can become invaluable to his company's planning effort by focusing it.

A company's focus may not be readily apparent. For instance, a company could continue for years making small automobile replacement parts without ever realizing that its real business is providing a service to individual car owners. Some companies, although they manufacture products, might find a more meaningful definition of their business to emphasize service. Perhaps a better characterization of the automobile parts company's business is that it is selling ease of access to its products. Strategic decisions not based on this important focusing definition could have disastrous consequences.

Thus, a firm's choice of focus colors not only its choice of product lines and markets it intends to pursue, but also how it intends to plan facilities. Observation of numerous manufacturing firms by the authors has revealed that the most common problem created by a poorly focused corporate philosophy is proliferation. The clear signs that a company lacks focus are proliferation of products, staff, and services. Although natural forces tend to encourage proliferation, only disciplined behavior can reverse it.

Part of the reason proliferation is natural is that marketing decisions often call for additional products and services, but only disciplined attention to costs and inventories will pare the list of older products. It is not uncommon to find numerous defunct items in a firm's storerooms and a matching set of products that are no longer sold but are maintained on the firm's product lists.

Pareto's Law of Income Distribution states that incomes are maldistributed among populations. Similarly, a Pareto principle applies to goods and services—only a small percentage of those provided by a given firm are sufficiently profitable to justify their existence. The error inherent in proliferation is that the company forces itself farther and farther away from what it has proven it can do well. This error is to be expected of a manufacturer that has a lack of strategic focus.

The ill effects of product proliferation are easy to describe but not necessarily easy to identify in action. A superfluous product with a low nominal profit margin can squeeze out a truly profitable product due to capacity limitations. The result is lower profit than is accessible without the superfluous product, a fact that makes its profitability negative. For example, consider two products that both require an hour's processing on a machine that is fully occupied every day. This machine is called a bottleneck. Assume that the first of the two products, Product A, has a profit margin of $300 per unit and that the second product, Product B, has a margin of $60. Assume further that either product can be readily sold in any quantities. Every decision to make and sell a unit of Product B will displace one unit of Product A, causing a decrease in profit of $240.

Companies make these kinds of decisions, forcing themselves into reduced profitability, and the cause is lack of focus. The usual sequence begins with manager discovering that he can market a product or service he does not now offer at an apparent profit. As the example above shows, only an accurate understanding of operational realities would show that a specific proliferation of activities is counterproductive. Thus, a correct focus of activities can be expected to result only if the company's strategic planning philosophy is sufficiently detailed to insist on fully informed bases for proliferation decisions.

A poorly focused philosophy affects diverse activities within the firm and limits its ability to compete. Recently, attention has been given to this problem and it has been labeled as a problem rooted in corporate culture, which is something like the ethos of the clan consisting of firm members. For example, consider a company that has always prided itself on its reputation for product innovation. If management were to suddenly issue its employees a directive saying that cost cutting was to become the company's priority, this ruling would create a proliferation of activities in just the same way as extending the product line would. This is because the corporate culture is difficult to change, and until it is changed employees will persist in following the same priorities they had absorbed before the new directive. Habits are hard to break. Proliferation results because people attempt to follow the new directive but do not divest themselves of ingrained attitudes readily.

The most common change in corporate culture that has taken place in manufacturing firms in recent years has been brought about by the computerization of processes. Consider the installation of a Material Requirements Planning (MRP) system. All procedures are affected by the new system, and it will only work if extensive changes are made in operational philosophy. Otherwise, employees will persist in performing their tasks in a haphazard mix of new automated procedures and old manual ones. If the staff was fully employed under the old system and is expected to be fully employed under the new, some tasks will receive inadequate attention. This is the classic example of proliferation due to lack of focus.

A change in corporate culture may be found to be a strategic necessity, but it must be effected as a change, not as a simple addition to the existing array of activities. MRP systems, when properly installed, can markedly reduce inventories and costs and can improve efficiency and productivity. Many such systems have been successfully installed, but reports on these cases always stress the large amount of education and training given employees to instill in them properly structured approaches to their jobs. Intelligently devised strategies, founded on accurate operational knowledge create these successes.

The more employees and managers diversify their activities, the more the individual activities compete with one another for their time and resources. Often companies have explicit policies that include a veritable laundry list of major objectives:

reduce costs
create new products
stress market development
enhance profitability
increase ROI
provide service

Such a list simply creates confusion because some objectives are contradictory and because they spread the attention of the firm over too many pursuits. For example, a plant manager may be called on to both limit costs and participate in new product development, both as high priorities. The usual consequence is that he acquires too many tasks to do any well.

The major objective of an organization is actually the one that receives the implicit emphasis. Many a manager gives lip service to cost reduction but only shows excitement in the face of product development. Eventually the message disseminates to all employees concerning true aims they should pursue, and the manager responsible for subverting his nominal purpose will be unlikely to realize that he has erred in any way. If, however, the firm's management were to argue through to an accurate statement of strategy, then the conflict between implicit and explicit objectives could be resolved.

A sharp strategic focus requires analytical study to devise it and continual objective observation to detect deviation from it. An intelligent focus for an organization may exist and only needs to be enforced, else it will have to be created initially, but focus must be consulted every time a strategic decision is made. This fact is especially important to strategies concerning acquisitions and expansion.

ACQUISITIONS AND EXPANSION

Many companies in recent years have become involved in acquiring other companies. For instance, U.S. Steel acquired Marathon Oil to broaden its product base within its capital structure and to make at least part of its holdings more profitable. To the extent of this objective, the acquisition has been a success. However, many companies have increased their debt-equity ratios beyond tenability as they have sought paper profit or immediate cash flow through acquisition. Unexpected market conditions left Dome Petroleum [50], for example, with an estimated 6:1 debt-equity ratio—an extreme case of overexpansion. Dome's debt service is now so expensive and its ability to obtain more capital so limited that it finds it difficult to perform its main function, the development of its oil field holdings. Numerous examples of wise and unwise acquisition policies can be cited.

There are sound reasons for purchasing another company. If an acquisition owns an existing distribution network paralleling the current needs of the acquiring firm, then that acquisition might be an intelligent move. However, many corporate takeovers are poorly conceived.

The reason mergers of companies (by whatever means) should be viewed warily derives from the need for focus. Mergers are prone to create a lack of focus, and the problem will afflict both entities in the deal. The acquiring firm will find itself with diminished management capabilities because the newly added division to the corporation diffuses management's attention. The purchased firm will find its own procedures disturbed by operating changes insisted on by the new parent corporation, impairing what was probably a successful operation before the merger. For example, it is common for the acquirer to find some of its own staff positions duplicated in the acquired firm. The redundant members of the new addition are usually fired, which immediately disrupts the staff of the acquisition. Cooperative mergers demonstrate much the same difficulties.

One problem encountered while trying to make mergers work is the lack of compatability of accounting procedures at the merged firms. In particular their cost accounting systems may be much different. However, this is a topic that is ordinarily considered within I.E. expertise. Thus, the planning of strategies that include acquisition should automatically call for help from I.E.'s. Since operational details are also important to how two merged companies will combine their efforts, I.E.'s should contribute their analyses of these details, too. If the inclusion of expert personnel in laying acquisition strategies were made a matter of course, the acquisitions would have a better chance of success.

The usual strategy that includes expansion of existing facilities or capacities derives from product positioning or product innovation. Almost the same caveats apply to planning for expansion as apply to acquisition. However, if the contemplated expansion falls within the corporate focus of activities, it could be a good move.

For example, a company may discover that its market area can be increased twofold by merely building a warehouse in Louisiana and stocking it from existing plants. If market studies and profitability studies support the idea, it may be a good one.

On the other hand, a plan may be proposed to expand facilities for the purpose of producing an entirely new product with which the company has no past experience. This idea is automatically suspect because it implies product proliferation at considerable capital expense. Even seemingly innocent expansion may be ill-advised.

For example, a particular company felt that its production facility was becoming capacity bound and was therefore considering the purchase of a new machine to relieve a plant bottleneck and increase its output. A study of the problem by the authors [42] showed that the plant was simply making the wrong mix of products, many of which were effectively unprofitable. They further demonstrated that an improved product mix would not only make more money, it would also make plant expansion unnecessary. The fact of plant operations that had not been well understood was that less profitable products were dominating the operating capacity of the bottleneck machine. Reducing the production of the low-profit products allowed higher-profit products to be produced up to the limits of market demand. The change in product mix obviated the need for the new machine.

Strategies that deal with increasing a company's capital base can be viewed as one class of strategies, while there are other strategies that primarily make demands on operating funds. One such strategic domain is research and development (R&D).

R&D

Competition tends to make any product or service obsolete by introducing new versions of it. A company can either ride trends and try to stay up with the opposition through imitation, or it can develop the new ideas itself. Ordinarily, new ideas will not be found unless they are sought; this requires investment in research and development (R&D). Yet few American companies commit more than a minimal percentage of revenues to R&D, while international competition stresses it and has made significant penetration into U.S. markets as a result. U.S. firms characteristically spend 1% to 5% of sales on R&D; Business Week [68] reports the American all-industry average during the 1983-84 reporting period to be 2.8% of sales. Unsurprisingly, the industries leading the way were software and hardware for information processing at 7.4% and 7.2%, respectively. Harder to understand are industries currently threatened by imports, such as fuel, steel, and textiles, which are investing less than 1% of sales in R&D.

It is possible in the short term to take an imitative approach to innovation; there are several hardware manufacturers in the computer hardware industry, such as Compaq and Amdahl, that do little but produce IBM-compatible equipment. However, one could argue that the original strategy of trying to undersell IBM at its own game was itself an innovation. No single firm can avoid being imitative from time to time, but, if it never creates new products or new approaches, it is difficult to see a prosperous future for it. The alternative to imitation is investment in formal R&D work.

R&D does not have to be done in-house, although it is easier to integrate in-house efforts into strategic activities. However, a lack of newly needed expertise or a rapid revision of market thrust may make outside help indispensible. One alternative would be cooperative research with universities. In some few cases work could be let out to independent laboratories.

Focus must again be considered when a company embarks on R&D. It will do little good to develop new products or services that lie far outside the competence of the firm's employees. One exception to this rule would be the case of a company developing an entirely new division to learn a new discipline from the ground up. If the strategic plan mandates the performance of work that requires personnel with currently unavailable skills, then the development of the necessary human resources must be made a part of that plan.

Competition is an excellent stimulus toward greater involvement in R&D. Even relatively stable products, i.e., those with long life cycles, will inspire some competitor to try to segment the market by introducing a minor innovation. A company that does not wish to see its market share stolen by competitors using this ploy must at least keep up with the changes. This means that some R&D effort must be conducted.

Responsibility for the cost of R&D is difficult to assign. This cost is often the responsibility of no one. Many companies have no formal R&D department and no specific R&D policy. By default product development is often proposed by Marketing and accomplished by a loose coordination of personnel in diverse departments. The authors have seen numerous instances when this approach has been attempted and has failed. A more organized system results under the aegis of a formal R&D department, coordinating the efforts of all professional efforts within the framework of the strategic plan.

Manufacturing engineers, design engineers, marketing experts, and industrial engineers are among those who might be included in the R&D effort. The role of the industrial engineers at minimum would include providing assistance in establishing costs, timetables, and goals and in ascertaining whether coordination between corporate systems can succeed in accomplishing R&D objectives.

FINANCE

A financial strategy is essential to budgeting, and any good accountant will declare that planning is budgeting. It is certain that any organizational unit not operating on a budget is highly suspect. How is its cost effectiveness to be measured? Its productivity? How is it to finance the elements of its strategic plan?

Budgeting must be done even when financing strategies are simple. A company could determine to spend a fixed amount of money out of sales each quarter and invest the remnant in a money market fund. That is simple. But such a policy takes no thought for tomorrow and contains no intimation of serious planning.

Finance theory indicates that most financing procedures require growth of the firm. Stock sales are based on the belief that the firm's asset base will expand. Financing of capital expansions by a company assumes that the ensuing generation of revenues should be sufficient to replace worn out equipment, cover operating costs, provide funds for investment in new opportunities, and repay the source of the financing. In nearly all cases these requirements have the effect of requiring growth, and growth takes planning.

Methods of financing will require a strategy. Should the company sell stock? If so, what kind? Common? Preferred? How about selling bonds? Perhaps it would simply be better to borrow or to mortgage existing facilities. Unfortunately, there is even theoretical debate about what mix of financing methods is best.

No matter what financing procedures are employed, how much money to obtain and how to budget will depend on two things: the schedule of cash outflows and the schedule of investments proposed by the long- range plans. Cash outflows are largely determined by costs of activities, and the understanding of most manufacturing costs is the special domain of the I.E.

In fact, accurate cost and price projection is an intrinsic part of growth, diversification, and expansion strategies. The better the accuracy of projections the smaller the funding uncertainties and the more intelligently finances can be obtained. For example, a company may determine to price its new product below cost to gain market share. It may sell at a marked loss, a condition that will require financing; this undercutting strategy must include a finance strategy as a natural component. Finance is also central to other kinds of short-term and long-term strategies, as will be seen.

SUPPLY SOURCES

Materials sourcing has grown in importance to competitive strategy. Japanese manufacturing firms tend toward using single suppliers where possible, receiving a given material from only one company, receiving deliveries as close as possible to when they are needed. In the U.S., firms have tended to avoid sole suppliers because they feared interdiction of their supplies by labor strikes or other interruptions. As a result, U.S. industrial companies have historically developed alternative suppliers, but there is considerable rethinking of this policy in America today. Perhaps the Japanese example has caused a reevaluation, or it could be that the tight economic conditions of recent years have made single suppliers more tenable.

For instance, the large U.S. automobile firms have limited the number of their suppliers to obtain faster turnaround of orders and better quality control of supplies. This trend is partly inspired by a desire for lower inventories and partly by consciousness of the costs of poor quality.

An alternative to using external suppliers is the development of in-house capacity for fabricating needed materials or producing subassemblies. This procedure is generically called vertical integration and is popular in the chemical industry. Ford Motor Company was pronouncedly integrated vertically at one time, although it is less so now.

Even if only a few items are produced in-house, they could provide considerable strategic flexibility. For instance, a company might operate a factory within the factory just to have access to a minimal supply of parts if its primary supply lines fail. Another common limited production strategy is to sustain a small amount of R&D work on crucial supplies as part of a search for better materials or fabricating procedures.

An extreme vertical integration strategy is the purchase of an entire plant of the kind that supplies needed parts. The drawback to such a plan is the usual one that resides with buying a business outside one's expertise. Exxon, as large as it is, found itself unable to succeed at acquiring and running a company, Reliance Electric, for electric motor development [64]. An interesting long-term acquisition strategy a company might follow has these steps: first, it develops a small in-house R&D and production facility to supply an important part in small quantities; then, as the plant continues to operate, both managers and workers acquire expertise in the disciplines attendant to the particular part's manufacture; lastly, it buys a company that makes the part, installs its own people in ancillary jobs at that plant until they know the new operation fully, and then eases out the old management.

The critical data that form the basis for supply decisions are of the kind developed by quality control personnel and industrial engineers. Scrap rates, results of acceptance sampling, delivery lead times, and material standards are the requisite numbers. From these numbers I.E.'s can prepare the cost projections by which alternatives can be compared. I.E.'s should also be called on throughout the developmental stages of major vertical integration strategies. Again, the I.E. expertise in projecting operating characteristics and costs of new facilties or expansions should provide the major factual basis for decision making.

DISTRIBUTION

Distribution problems are abstractly similar to supply problems and can similarly be accomplished in a variety of ways, including doing it all in-house, farming it out to wholesalers, or using wholly owned retail outlets. Transportation modes can be chosen from diverse alternatives including company trucks, contracted truckers, rail carriers, barges, and ships.

Distribution decisions affect capital investment,speed of delivery of products to purchasers, wholesale-retail policy, profits, and the ability to control costs. The choice of a distribution network may also dictate the market area. Product availability and pricing decisions are so heavily influenced by distribution decisions that they become more important than is generally recognized.

Distribution analysis and implementation is a good example of a category of strategy for which I.E.'s are especially well suited. The requisite data base on which to base rational distribution decisions includes lead times, unit costs, and capacities. Lead times must be understood for each type of carrier and each route. Costs of shipping a unit of cargo via each route and mode of transit must be ferreted out. Capacities of carriers and warehouses at both ends of shipping routes must be included in the data used to develop strategic distribution plans. These are the kinds of information I.E.'s naturally understand and pursue well.

MARKETING AND APPROACH TO COMPETITION

It can be argued that marketing strategies are naturally of the short-term variety, and it is true that many marketing decisions deal with strictly short-term approaches to increasing sales volume. However, once high-level policies aimed at either enhancing profitability or lowering costs have been determined, the resulting long-term marketing strategies must be readily available for consultation. A company cannot afford to simply react to every change in the competitive environment without the guidance of general principles and objectives.

A company must decide whether it is selling to commodity consumers or to a specialty market or to some mix of markets. If conditions seem to require direct confrontation with competitors, standard policies on how to conduct an aggressive strategy must be established. For example, will the company conduct promotions? Will it cut prices? If so, how and why and when and where? What articulated strategy guides market expansion? How are new product introductions to be accomplished and why?

It might be said that each of the above decisions can be made on the spur of the moment, but more commonly such decisions are made amid confusion too slowly to capture the moment. Only existing, comprehensive statements of marketing strategies can provide the framework for rapid, intelligent decisions to be made with neither inordinate fearfulness nor overconfidence.

One important long-term strategy decision is how to approach competitive pricing. An excellent example of pricing is seen in gasoline sales. A gasoline distributor that runs retail outlets may find himself confronted with price-cutting by a competitor, but he cannot respond in kind without losing money, because profit margins on gasoline sales are typically low, even at usual prices. A policy suddenly constructed on the spot could be disastrous. A long-term strategy derived from a standing policy could be disastrous, too, but it is less likely that cohesive ideas, pieced together dispassionately, will lead one astray.

The most important elements of data needed to select price strategies are accurate estimates of product costs. The I.E. is the professional that should be involved in developing the data, but he would be even more effective if he were fully involved in the thinking that goes into selection of strategy.

As part of the support for strategies that rely on promotions of products, the I.E. would also need to be able to provide estimates of production and distribution capacities that could be applied to providing the rush of units needed to make promotions successful at creating enhanced sales. Knowledge of the strategic alternatives under review would help the I.E. in his search for truly useful information.

ORGANIZATIONAL STRUCTURE

How a firm is organized is undeniably long-term in its effects, because corporate structures acquire inertia opposing change from inception. The reason for this fact is readily explainable. Most employees of any organization feel comfortable when they know their jobs and how success is measured. They feel personally threatened whenever marked changes in the definitions of jobs and expectations are proposed. Most people fear that organizational changes will affect them adversely. Thus, they prefer that any changes made are not sweeping. This attitude creates an attachment to habit, and habit is the most common corporate motivation. Once instituted, boxes in the organization chart are difficult to eliminate or redefine. A company that realizes this verity will plan to maintain as spare a structure as possible.

A given company's structure is usually represented by its organization chart (see Figure 1.1). There are two modes of classification on such a chart: hierarchical level and function. Horizontal divisions depict separation of function; for example, consider the separation between manufacturing and sales. The vertical divisions depict decision making levels and levels of rank. Line activities extend up and down and include end uses, such as sales and manufacturing. Staff functions are usually responsible only to higher levels of personnel, either to the vice president or president level. In the organization chart of Figure 1.1, Accounting, Finance, Management Science, and Computer Science are shown as staff functions.

Temptations that are best resisted are the creation of new staff departments, the addition of a new layer of decision making, and the creation of additional corporate divisions. A paraphrase of the second law of thermodynamics describes the situation: every decision to expand the organization chart increases confusion.

There are numerous organizational policy decisions that require broad, strategic guidelines.

- Treatment of corporate divisions:
  - The efficiency of a division is seriously affected if it is moved from line status to staff status. Policies should preclude such moves. Particular divisions that have their efficacy damaged by such a move are engineering and R&D.

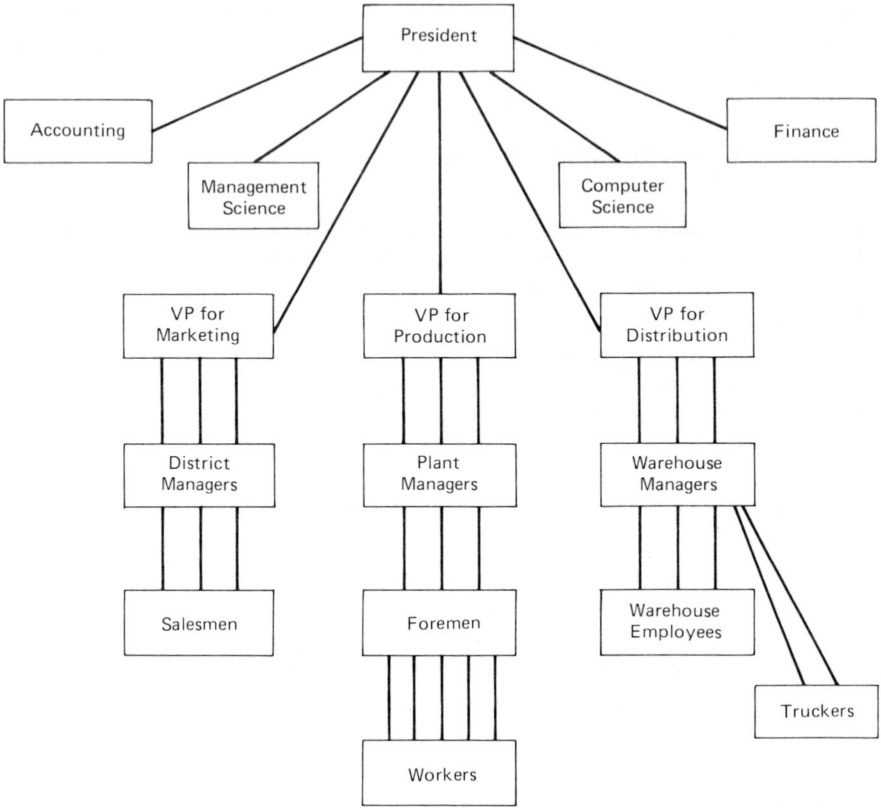

Figure 1.1. Organization Chart.

- Accounting policies:
  - Decisions must be made from time to time whether a division should be a cost center or a profit center. If cost centers act as suppliers to other divisions, then transfer pricing policies must be rationally developed.
- Accountability:
  - Every individual who is insulated from accountability for profits or costs weakens the effectiveness of the firm.

I.E.'s should always be involved in questions of organization, if for no reason than the generally acknowledged fact that they are the productivity experts. It is they who should manage productivity measurement systems, including the study of white collar productivity. At present, few formal systems that appear to have an engineering basis are applied to white collar productivity. The dominant measures are the firm's quarterly ROI and behavioral expectation scales. The I.E. must bring consciousness of engineering approaches to the evaluation of productivity within organizational structure—he must be the motive force.

I.E. divisions are, unfortunately, often moved to staff level. This severely weakens their effectiveness, removing them from direct responsibilities within the production function. Yet it is I.E.'s who can best describe the job flows and systematic interactions on which the shifting of divisions should be based.

The role of the I.E. in the study of whether an organizational unit should be a cost center is readily evident, again due to the I.E.'s reputation in cost analysis. This decision is a strategic one, but it should be based on the factual evidence that I.E.'s can best provide, including proper transfer pricing levels.

The integrated role of blending analyses of these important organizational problems into his domain will require determined and forceful I.E.'s. However, the modern role of the I.E. should extend his involvement into those strategic aspects of management that can clearly benefit from his expertise.

THE ROLE OF THE INDUSTRIAL ENGINEER IN LONG-TERM PLANNING

While the dimensions of long-term planning discussed above are perhaps familiar to many industrial engineers, the issue that must be addressed is how they bear on the practice of industrial engineering. Typically the industrial engineering approach to most of the issues discussed here is indifference. The I.E. may recognize the absurdity of expanding the product line at an outmoded facility but rarely does anything about it. No attempt is made by the I.E. to cost out the proliferation or to look at the opportunity costs of failing to replace out-of-date production equipment except when he is requested to do so by higher management. Why is this the case?

There is almost no industrial engineering guidance provided to the research and development activities of a company. New products and manufacturing processes need the expertise of the industrial engineer at the planning stage, not after investments in them have reached the point of no return. Why this oversight by management?

The industrial engineering discipline is concerned with costs and with resource utilization. Why, therefore, is the I.E. not consulted on pricing decisions and finance strategies? The I.E. again is involved only after the fact, not as a full participant of the decision making process. Why?

Decisions to vertically integrate a business have marked influence on the conduct of manufacturing and industrial engineering. The same is true of distribution decisions. Again, the I.E. is involved only after the fact and not during the decision making process. Why?

The reasons for the failure of the industrial engineer to become involved in essential features of long-range planning are many and varied. I.E.'s have not promoted their talents. Some have been content to accept the limited role many companies have assigned them. The companies themselves have characteristically undervalued and misunderstood the I.E. function, also. In many I.E. departments it is not uncommon to find non-I.E.'s performing the simple tasks that are labeled as industrial engineering—tasks such as the maintenance of time standards.

However, the training of I.E.'s is broad, ranging from the study of costs in manufacturing to data analysis, mathematical modeling, production control, and wages and incentives. They are trained to effectively learn the total operations of the manufacturing environment. If an I.E. were to also learn the concepts and purposes that motivate strategic planning, he would be formidably equipped to manage a holistic view of major long-range decisions.

I.E.'s have been somewhat more involved in short-term planning but not to the degree they should be. As in long-term planning the I.E. needs to educate himself to strategic goals and to aggressively pursue the opportunities for profitability that he can best exploit due to his unique training. Let us now consider what aspects of strategic planning demand short-term abilities, plans, and facilities.

2
SHORT-TERM PLANNING

Within the broad framework of long-term policies aimed at enhancing profitability or reducing costs, there is a range of short-term issues that must be addressed from a base of solid management principles. These are the problems of short-term strategic planning. Even after long-term strategies are determined, there are constantly changing conditions that must be reacted to or anticipated. Much as a football team's defensive strategy may be geared to stop the run, a company's long-term strategies may be geared to enhancing profitability. However, as each play develops in football and in business the strategy must be translated into immediate application. For this reason, a company trying to plan future strategies must adopt rational operating plans for their short-term implementation. An important example of this short-term requirement can be seen in the competitive positioning of products.

PRODUCT POSITIONING

Product positioning requires strategies aimed at improving market position of an existing product. Positioning strategies contain unavoidably short-term components because they must adapt in reaction to changing conditions that can be anticipated only approximately in the long term. Competitor pricing, economic climate, market maturity, product life cycle, costs, new product introduction, and interest rates are some factors that require short-term positioning strategies.

If a company's main competitor in a given market line inaugurates a price-cutting campaign, then a response is required. A price war might be the result. A company sure of its stronger financial position might readily enter into a price war with the cheerful hope that its competitor will starve on its self-imposed, small earnings. Another response would be to increase advertising expenditures and deliver the message that, although some products are cheaper, OURS is of higher quality; since the breakup of AT&T's telephone monopoly, this has often been the tenor of the AT&T long-distance commercials. Establishing a preeminent role of one's product as the industry's high quality price leader is a not infrequent positioning ploy; this is also the approach taken by Hallmark Cards.

Economic vicissitudes can require short-term adaptation of plans. For instance, interest rate increases can inhibit a company's zeal for making rapid product deliveries because quick response can be achieved only by maintaining expensive product inventories. Similarly, general economic conditions can affect the ways products are marketed or advertised.

If a product is approaching an expected decline in its profitability, the company may begin to reduce its support for and promotion of that product. Advertising, sales campaigns, and even product inventories may be reduced near the end of its life cycle. A stagnant or declining market for a product would also signal a company to withdraw from the activities that promote its sale.

These and other positioning situations should be dealt with according to the general principles of company policy and long-term strategy. Policies may include prohibitions against excess inventory, lower limits on acceptable product returns, or strict cutoffs on the duration of a product's offering. Many such policies require quick response to ephemeral conditions, and short-term strategies establish the firm's ability to react. Rational short-term strategies result from the application of sound general principles; thus the frequent need for implementing short-term strategies emphasizes the need for intelligent long-term approaches to product positioning and to all the other long-term strategies with important short-term aspects that are discussed below.

PRODUCT DEVELOPMENT

New products come to life for diverse reasons. Some spring forth from technological innovations, as the home computer has derived naturally from silicon chip technology. Some products are introduced in an attempt to segment a mature market. Sometimes a genuine, unfilled need is detected, and the product to satisfy the need follows. Product improvements, such as fuel-efficient cars, led to the birth of new products, such as smaller cars. There are also those cases such as that of the hula hoop in which a product is introduced even though the need for it is not evident.

A manufacturer of goods or a provider of services needs a long-term product development policy dictated by its long-term strategic focus, but the actual work of product R&D and product introduction is a short-term strategic problem. Consumer need for a product often comes into existence without fanfare, and competitors frequently try to conceal their new product plans or advertising campaigns until the last minute, so only an absurdly complex long-range plan could contain the detail needed for rapid response to all contingencies. Workable, general, short-term plans must be constantly available for application.

Similarly, product innovation may result from a sudden design breakthrough, again pointing up the intrinsically short lead times characteristic of product development. This again points to the need for the firm to have ready patterns of action installed in its corporate culture.

Short-term decisions, if they are to be effective, must be based on realistic assessments of alternatives and on accurate information. For instance, a decision to rapidly increase a product's output must be founded on the information that the increase is feasible. Further, this short-term result of strategic principles should also be based on cost information that will support this approach to profitability enhancement. Ideally, the strategic plans would provide for the computational tools that allow for the rapid, but correct, assessment of such short-term activities.

Computational tools in turn require up-to-date data bases containing accurate cost and capacity data. For example, an auto manufacturer must understand the costs and lead times associated with die changes before it can plan body design changes intelligently. Changes in work methods must be costed and have time standards associated with them before meaningful operational plans can be laid. The repercussions of substituting forgings for castings must be translated into times and costs. And all of the information on alternatives must be developed into sequences and schedules, capacity utilization analyses, and implied product mixes before feasibility and economic impact of product/process changes can be rationally studied.

Once again it appears that the I.E. should be strongly involved in product development strategy due to his ready understanding of process and his training in integrating concepts into holistic viewpoints. Numbers may be generated by foundrymen, manufacturing engineers, marketing experts, and I.E.'s, but it is the I.E. who is by training and experience best prepared to compose the big picture from the components. The I.E. is also the most reasonable choice for the person to design analytical and monitoring procedures for studying proposed and ongoing strategic implementations for the same reasons. Monitoring will be seen to be an exceptionally important short-term aspect of strategic plans, as, for example, in the tracking of product life cycles.

PRODUCT LIFE

A product's life cycle may actually be planned from birth to death. For instance, a razor products company may have an entire string of products planned to succeed each other at two year intervals. However, many products have unpredictable life spans, determined by a variety of factors, including the actions of competitors and economic conditions.

A typical product life cycle looks like the graph shown in Figure 2.1. The product begins by requiring development funds, eventually breaks even, begins to become profitable, grows, attracts competition, and declines. There are ex-

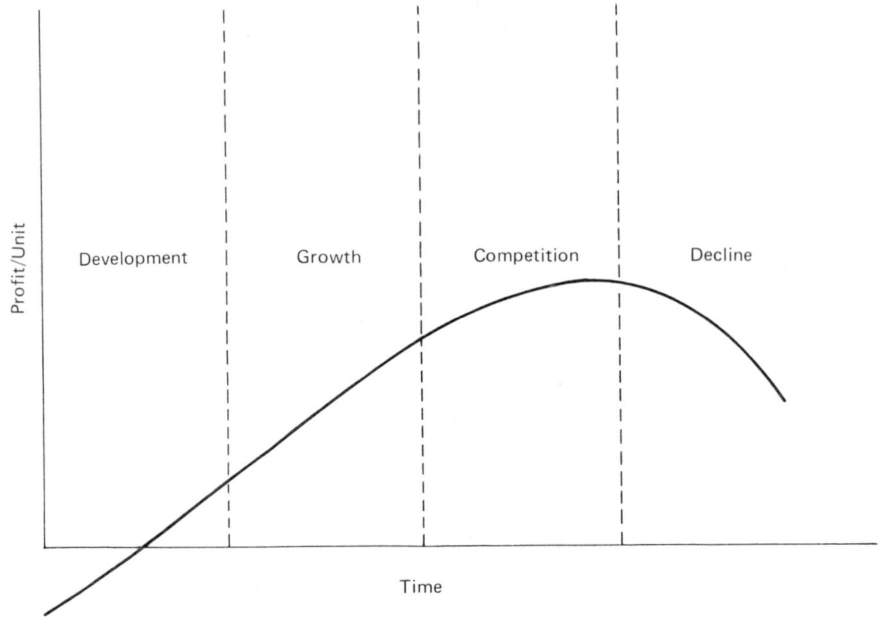

Figure 2.1. Product Life Cycle.

ceptions to this pattern, but it is common, and it serves as a frequent basis for planning the strategic handling of products.

Thoughtful product life planning requires a foundation of established strategic principle along with artful short-term implementation procedures. At the very least there must be methods for tracking a product's life cycle and estimating which phase of life it is in. If its obsolescence has been planned, then the procedures for phasing out the product will have been be determined beforehand. Further, the conditions that trigger the phaseout will have been specified.

The monitoring aspect of product tracking is important not only to the handling of those products that follow common life cycle histories but also to those that become exceptions and those that are expected to be exceptions at their origins. Sales assuredly must be tracked but their observation must also be associated with production information. For example, how have methods and cost standards changed for the product? How much lead time is advisable to cut the costs of phaseout, especially as it affects inventory costs and capacity utilization? These are questions that are best coordinated between marketing personnel, I.E.'s entrusted with the monitoring responsibility, and a clear statement of strategic plans.

Some few products defy expectations. Two opposing examples can be found in a single company—Coleco. Its Adam* home computer was expected to do well in the intermediate-price home market; it did not. Its Cabbage Patch Kids* were expected to excite limited Christmas sales; they became the season's hottest item. These two cases represent the poles that monitoring and strategy must attempt to span. When a product dies unexpectedly, rapid cancellation of materials for it should be activated. Similarly, liquidation of remaining finished goods should be conducted expeditiously. When a product excites sudden, large demand, profits can be seized from this serendipity only if organized plans for boosting production and distribution are readily implementable. The implied need for the integration of product monitoring and contingency planning for production suggests the need for I.E. involvement as a principal participant.

High-tech products are especially volatile, so clearly thought out life cycle strategies must be developed for them, especially their phaseout. The danger inherent in product discontinuation is that it will be done too late or in the wrong sequence. Consider the case history of a company making computers for automotive testing. It had planned to discontinue an old product at a given date, at which time it would commence production of a new design. Unfortunately, parts inventories were ordered by an automatic, computer-forecasting MRP package because no one told the computer of the discontinuation planned for the old product. By the time a human checked the warehouse, parts sufficient for 500 of the old units were beginning to gather dust. The unavoidable, expensive solution to this problem was to make the 500 units of the old design, sell them for whatever they would bring, and then switch to the new product. Compounding the unpleasantness of this solution was the fact that making and selling the old products reduced demand for the new product by 500 units.

Product life must include detailed checklists of procedures and schedules, such as when to revise the MRP program, the maintenance and supervision of which is best left to I.E.'s. Scheduling, sequencing, methods analysis, cost analysis and monitoring, product mix studies, and algorithmic integration of processes are the operational domains in which I.E.'s excel. Thus the organization and operation of the monitoring and preparation aspects of product life planning should prominently feature the work of the industrial engineer.

PRODUCT MIX

The broad policies that determine a company's product mix should dictate its strategic focus, but in the short term specific product mix questions must be answered, especially questions related to proliferation. A basic problem that all

* Adam and Cabbage Patch Kids are registered trademarks of Coleco.

organizations face is proliferation. Except in the case of serious financial hardship, organizations tend to get bigger. More managers are hired; more workers are hired; more buildings go up; more services are provided; and more products are manufactured. Preventing unwise proliferation is, therefore, a constant challenge.

A major step in avoiding unwise proliferation of products is determining what constitutes a rational product mix. Product mix analysis is logically an integral part of aggregate planning, but it is also a distinct discipline. The criterion guiding the selection of a good product mix is that it should be highly profitable. To be profitable it must consist of products chosen from two categories: those that have high (or potentially high) profit margins and those with lower profit margins but high customer demand.

Some companies prosper by selling only one of the two categories of products. There are specialty companies, such as a bathroom fixtures dealer than sells only gold fittings. More commonly, a specialty line will be only one of several lines manufactured by the company. High-priced items naturally attract potential competitors, so only those specialty items that are difficult to imitate or that have strong name identification are likely to remain one company's private product.

High-volume products ordinarily attract heavy competition, also. It is difficult to imagine one company holding a monopoly on a rapidly selling item. Numerous competitors fighting for the same market usually cut prices, so it is rare to find a commodity item with a large profit margin. Nonetheless, high-volume items are the mainstay of most manufacturers.

Although conditions may dictate that a portion of production be stable and predictable, the uncertain portion must be planned in the short term. The current product mix a company aims to sell is a function of numerous factors, many of which change daily. The mix of orders received or the availability of materials may require frequent juggling of priorities. These necessities in turn require reactive capability.

The ability to react rapidly to changing conditions is a function of preparation and firm knowledge of purpose. The strategic plan lays out the general course, including assignment of roles to individuals and to corporate disciplines. As has already been discussed, the role of the I.E. is central to any workable plan to restructure sequences, plant loadings, materials orders, schedules, and inventories. This fact makes the I.E. an important cog in product mix planning.

METHODS OF PRODUCTION

Long-term capital investment strategies should establish most of the equipment configuration and the layout of production facilities in a manufacturing firm.

However, there are still decisions to be made in the short term as to how to use facilities to the best advantage.

Consider the problem often encountered in which a variety of products share the same pieces of equipment during their manufacture. In a graphite electrode plant the electrodes of all sizes are heat-treated in a limited number of autoclaves. In a rolling mill products of various gauges are rolled on the same equipment. A foundry melts a variety of alloys in the same electric furnace. A print shop prints a number of different posters on the same machine. Even service divisions must share people between jobs.

In all these cases the question that arises is whether a change in production methods might relieve the bottlenecks. Sometimes the problem cannot be resolved. Each electrode simply has to serve time in an autoclave. In the rolling mill, though, astute study might show that an idle set of rolls could be used to relieve a limited amount of the gauge reduction backlog. An idle five-color press might similarly relieve some of the pressure on the three-color printing press.

Sometimes ingenuity must be employed to get a job done at all. A plant with a machine down might send unfinished products to another plant for completion in a pinch. One plant that grinds material discovered it could substitute one grinder for another in emergencies. Another plant in straits might discover it could substitute materials or personnel to relieve shortages.

The need for solutions to production constraints is not the only reason decisions about production methods are required. At times different production procedures must be considered and the better chosen. For instance, a metal products plant may need to choose between casting or forging to produce a desired new piece. Comparision of the alternatives in this case, or in any of those cited above, must be based on accurate data to be effective.

All these decisions should be based on whether the alternatives are possible, whether they are advisable, and whether the costs involved are acceptable. It would seem natural that I.E.'s should be very active in answering these questions in light of existing strategies. When it comes to questions of how people should perform work, the I.E. is naturally called upon first, but his combined understanding of manufacturing and cost control should put him in the forefront of most methods-of-production decisions.

INVENTORY PLANNING

Broad inventory guidelines should derive from long-term strategies, but in the short term inventory is intimately connected with production planning. Both the production control and cost control functions interact with inventory planning and control. Two dissimilar requirements must be met—inventories must be available when needed, but excess stocks must be avoided.

Inventories consist of raw materials (including purchased parts), work in process (WIP), and finished goods that have not been sold. Inventory planning in the broadest sense should aim at assuring that needed parts are on hand in a timely fashion, arranging for subassemblies and fabricated materials to be where they are needed on time, and providing quantities of finished goods sufficient for safety stock criteria. In addition, inventory planning should seek to avoid needless surpluses of inventories at each stage.

The Japanese kanban system is an interesting example of an inventory policy that arranges for timely flows of materials while virtually eliminating buffer stocks and finished goods inventory. This system is also called "just-in-time" production. As a paradigm for a specific type of operation, one that can assume fixed demand over time, kanban is excellent, but not all facilities fit the fixed-demand assumption, so plant policies must usually be determined plant by plant.

It is extremely important to limit inventories in periods of high inflation, because inflation leads to high interest costs for borrowed funds, which make inventory holding costs dominated by the continuing costs of their financing. Even though long-term strategies may lay out general inventory plans, short-term operational considerations require that the technical activities of inventory planning and control be performed at the shop floor level.

Again, this seems a natural effort for an I.E. to oversee, especially now that technically based procedures, such as kanban, are more commonly employed. Inventory planning and control is a large part of an I.E.'s formal study. He is educated to inventory control as a distinct discipline and to its integration in production planning and control. He understands the cost accounting aspects of the topic and its influence on scheduling. Since I.E.'s are trained in the uses of mathematical modeling and computer-assisted manufacturing, they readily grasp the purposes and procedures of computerized manufacturing systems, and they are probably primarily responsible for deriving the databases of installed computerized systems. I.E.'s are the central personnel in installing computerized, integrated manufacturing systems. I.E.'s therefore seem irreplaceable in the process of implementing strategic inventory policies.

SCHEDULING METHODS

It can be argued that scheduling is the most important operational planning function in industry. It has been argued that most problems encountered in production spring from poor scheduling. Poor scheduling in a plant is evidenced by workers frequently waiting for parts or by wildly varying buffer stocks between stations. The former demonstrates poorly understood lead times; the later shows poor line balancing. It is easy to identify a plant with poor scheduling.

A poorly scheduled shop usually exhibits a high degree of expediting—people darting hither and yon trying to hurry along sadly overdue jobs. A large amount of rework and poor-quality output are signs of poor scheduling. It is difficult to perform badly scheduled work. There is a common saying in many inefficiently run shops: "We never have time to do it right, but we always have time to do it over."

The failures of technical scheduling systems, such as MRP, are commonly due to poorly developed data upon which schedules must be based. The requisite data include machine and personnel capacities, labor standards, lead times, material standards, machine standards, and flow rates. Only after accurate numbers of these kinds have been developed does scheduling, by itself or integrated with other functions, have a chance to succeed.

A good I.E. can obtain the needed data and derive the kinds of scheduling strategies that will work. Schedules can be produced that will allow work to be scheduled in an understandable sequence with adequate safeguards for quality control in a cost effective manner. This is exactly the type of function for which an I.E. is trained.

DISTRIBUTION PROCEDURES

Distribution consists of not only shipping product but also of keeping track of when product is ready to go and whether it reaches its destinations. Operational departments affected by distribution strategies include Traffic and Transportation. These strategies also affect Marketing and may determine the effectiveness of sales order entry and release systems. Corporate production control can even be affected by distribution.

As simple as it may seem, it is often difficult for a company to find an order that its records say it has completed. At some of the more primitive steel mills innumerable slabs of steel have been known to disappear; simultaneously, a considerable pile of dead inventory whose purpose has been lost has sprung up. Some rolling mills deposit rolled coils of metal on the plant floor in such a helter-skelter fashion that product gets lost. Some manufacturing plants lose partially finished orders and eventually lose the sale when the customer becomes impatient and cancels his order. Sometimes full orders are shipped to the wrong customer.

These problems undermine strategic plans and reduce profitability. Systematic procedures, based upon a firm understanding of how distribution works and how to improve it can readily be produced by I.E.'s, whose training ideally equips them to study flows and procedures and then streamline them. Lack of a rational system is the usual root cause of incoherent distribution activities, and this rationality the I.E. can provide.

Systematic handling of storage and shipping is imperative if profitability is to be enhanced. If every step in a manufacturing process is performed effectively but distribution is not, then profits can disappear in that one step. Solutions to distribution problems may not be complex but they are likely to require the organized approach of the trained I.E. One I.E., finding that his rolling mill was losing track of orders on the shop floor, inaugurated a painting project. He divided the storage areas into three portions based upon the profitability and volume of product lines. Then he further divided these three areas into gauge ranges. He also installed a procedure by which each coil delivered to an area was tagged by the fork lift driver who brought it. Each tag contained the alloy serial number, the customer serial number, the order number, the date, the coil gauge, and the driver's name. Coils were regularly checked by the shop foreman, shipping personnel, and marketing personnel. If a coil was found to be misplaced, the driver who put it in the wrong sector was chastised; if shipping was slow in sending out a completed order, that department was chastised; and if marketing found that the lapse between a coil's completion and the production of the remnant of its order was excessive, the production people would hear a complaint from marketing management.

In any high-volume operation some orders will be lost, misplaced, or sent to the wrong place, but intelligent distribution strategies can provide the climate that will allow the operational disciplines to mesh more effectively. The ideal personnel to bring to this problem are I.E.'s. They are devoted to promoting efficiency and profitability and can thus serve as objective interlocutors between departments, including Marketing.

MARKETING PROCEDURES

Marketing is a discipline not commonly viewed as lying within the expertise of I.E.'s, but this is not to say that I.E.'s cannot make a contribution to marketing strategy. One of the most common sources of loss of profit derives from poorly formed strategies on the part of marketing personnel. Understandably, salesmen want to sell, and they want to satisfy the customer's needs as rapidly as possible, but problems can be created by overly zealous salesmen, resulting in unnecessary costs.

The usual salesman's contention is that he is losing orders because products are not in inventory whenever his customers need them. If this complaint is translated into production strategy, then an inordinate quantity of finished goods inventory may be created. In addition, if the company seriously tries to guard against possible shortages of every product, then production will be constantly diverted toward short, inefficient runs, simply for the purpose of restocking arbitrarily dictated products.

Some companies have allowed their marketing departments to attain undue influence over production to the degree that marketing can literally demand the production of rush orders. No long-term production strategies can be effected in the face of such unpredictable interruptions of production schedules, and cost/unit guidelines cannot be satisfied. This problem is yet another example of the product/process dichotomy of viewpoints that frequently come into conflict in manufacturing firms.

The role of the I.E. in marketing needs to be that of an educator. He needs to teach the sales people where costs are incurred and how cost-effective production strategies should be followed. He must further educate Marketing to consider the possibilities of product substitution and to avoid selling certain products altogether. Also, salesmen must be informed as to the most cost-effective ways to write orders and the most reasonable lead times to promise customers. In general, he needs to dissuade marketing personnel from doing the counter-productive deeds they may do thoughtlessly.

In an intelligently structured marketing strategy, information must be kept current so that the short-term strategies can be adapted. Standard flows of information between marketing and production need to be established so the departments can coordinate with each other in a cohesive approach to improving operations. I.E.'s can design the requisite information flows.

It must be noted, however, that this aspect of integrated manufacturing planning is not often given a great deal of attention. Yet, an I.E. possessed of an understanding of strategic goals and a holistic overview of corporate operations will surely perceive Marketing as simply one component among many to be integrated into a harmonious whole. He is one of the few people within a firm that may realize the dual importance of the product and process emphases and how they interact with facilities utilization.

FACILITIES UTILIZATION

If one's firm has a set of fixed facilities—buildings, equipment, etc., one might take the view that the only detail remaining toward the implementation of production plans are operational details. However, there are a number of intervening strategic facilities issues to be decided before operational plans can be developed. A good example of strategies for operations is provided by maintenance.

For example, given that equipment and facilities are to be maintained, what is the best strategy for accomplishing maintenance? Should there be an assignment of priorities made to each type of maintenance, or should maintenance jobs be assigned as the need for them becomes apparent? Should there be a regular schedule of preventive maintenance? If so, how is this schedule to be derived? What measures of effectiveness should be used to evaluate the maintenance procedures and personnel? Maintenance is only one aspect of facilities utilization that requires the development of coherent operational strategies. Another category is materials handling. Another is inventory picking. Another is traffic. The common thread between all these problems is that they deal with management of operations within a fixed facility.

These are topics strongly developed in I.E. collegiate curricula. I.E.'s learn the probabilistic nature of machine failures, for example, and they learn that there are cost tradeoffs between capital and operating expenses. Their resulting awareness allows them to readily understand the cost tradeoffs upon which preventive maintenance and equipment replacement decisions should be based.

Industrial engineers also study the problem of choosing an effective plant layout in school. This study requires the understanding of flows, locations, and sequencing of operations within a fixed facility. The understanding of operational flows and layout make the I.E. well prepared for solving such problems as equipment placement, production line design, warehouse layout and picking procedure specification, materials handling methods and flows, and traffic flows.

Coupled with the I.E.'s sensitivity to cost minimization, efficiency, and integration of functions, he is the ideal individual to coordinate the preparation of facilities utilization plans and their meshing with the integrated strategic plan. His computational training and integrative insights also prepare him well for dealing with the growing modern problem of designing, installing, tailoring, coordinating, and monitoring computer integrated manufacturing systems (CIMS).

COMPUTER INTEGRATED SYSTEMS

Probably the most comprehensive efforts in developing tracking, reporting, and control procedures have been made by the Japanese, and these efforts descend from a pervasive aversion to inventories, exacerbated in recent years by high inventory holding costs. The most thorough system, the kanban, is not even a computer-driven system. All operations are tracked in small batches by use of batch cards, and the location and status of the cards constitute the entire tracking, reporting, and control system. However, one interpretation of the kanban system's rationale is that it aims at achieving just-in-time production.

Just-in-time operations suggest a conscious minimization of inventories throughout the production stages, and this in turns implies stringent tracking and control of operations. To many professionals concerned with inventories in operating environments, such control appears to be available only in the context of computer control. Thus, one of the most popular topics among production personnel in America has become that of computer integrated manufacturing systems.

The concept is appealing, and the resulting cost savings and efficiencies from computerization can be impressive, but there are a number of difficulties attendant even to small-scale computerizations of manufacturing systems. Whether to computerize an operation is a question that should be explored in detail prior to making the decision. Often the computer system adopted is too slow or inaccurate to be effective in a manufacturing environment. Often the computer software is not adaptable to the job at hand. In other cases a computer control system is indispensible.

In either case, information flows must be of the sort the system can use effectively. Any system that does not allow for the tracking of jobs and monitoring of schedules is not very useful. Unless an information system can provide such data in a timely fashion, it cannot be useful in production planning and control. An I.E. studying the potential usefulness of a computerizing system will be interested in both the flexibility of the system and its direct and indirect costs.

The I.E. will see the adoption of a CIMS, or of any smaller version of such a system, as a long-term commitment on a par with facility purchase decisions. Once such an expensive procedure is installed it is difficult for any manager to declare it an expensive mistake and discard it. In addition, the I.E. knows that a manufacturing firm tends to adapt to the computer's procedure rather than vice versa.

Successful computerized systems have been installed, but these cases invariably require extensive training of personnel and tailoring of procedures. Costs of these installation activities are ordinarily a high percentage of the total initial costs of the system [26]. Since I.E.'s have considerable experience and training in the diversity of topics, including costs, computerization, production planning and control, inventories, and scheduling, that the CIMS tools address, a logical approach to analyzing a contemplated adoption of a computer system is to simply assign an I.E. to its study.

The questions he will try to answer are:

Can it work in our operating system?
Is it cost-effective?
How will it work in our environment?
What tailoring, training, and adaptation will it require?

Sometimes an I.E. studying a less-than-optimal computer information system will find that it misfires only in a few instances while remaining useful for other functions. In this case the I.E. might be moved to adopt a strategy of using the studied system but of also installing special-purpose information systems to handle the exceptions.

On the other hand, the I.E. might determine that the computer system is useful but that it must be augmented by a range of additional procedures to tailor it properly to its operating environment. For instance, manual reporting systems can be used in concert with large software packages to fill in the gaps.

There are many alternatives to computerization short of buying a complete, expensive CIMS package. A good I.E. would naturally consider a number of these alternatives. In some cases a good strategy may be to purchase a microcomputer and assign it one restricted task it will perform over and over. It may be a good strategy to develop a small interactive program as a decision support system for scheduling or inventory decisons. An open and flexible mind will be required for the selection of utilitarian strategies. An I.E., by training and experience, has the credentials to provide just such a mind.

THE I.E. APPROACH

Much of what an I.E. is called upon to do in the short term falls under the category of "sideways thinking," i.e., coming up with ideas that may not have been tried out before. The best sideways thinkers are those adept at viewing problems in a holistic fashion instead of only one piece at a time. A good I.E. has the academic training to provide a fairly broad view of his operating environment when hired from college. Usually, he also works on the plant floor, labors in a line capacity, and acquires a feel for manufacturing realities.

The additional piece of understanding he needs to possess an understanding of the firm's big picture is knowledge of its strategic philosophy and plans. Given this insight into the overall corporate focus of direction, he will be in a position to incorporate his knowledge of functional reality into the strategic vision of the firm's goal and objectives. He should then be in a position to take his logical place as an active member in the strategic planning and implementation process.

3
THE FAILURES OF STRATEGIC PLANNING

The authors have encountered many instances of corporate attempts at strategic planning and have partaken in both the planning and implementation of such plans. Often these plans have been unsuccessful in achieving their nominal goals, largely for those reasons already suggested: lack of focus; lack of underlying reality; and lack of experience on the part of the decision makers.

The lack of experience is often the reason for the other two problems, so let us consider this factor first. This roadblock to successful strategic planning is usually exemplified by a "tunnel vision" on the part of decision makers who are experienced in only one facet of the business. Finance people may perceive the firm's best opportunities to lie with acquisitions, seeing the basic business as little more than a holding company for financial manipulation. Marketing personnel are likely to think strictly in "product" terms to the exclusion of "process," while production personnel will think just the opposite. These diverse, limited viewpoints become clear even to company outsiders who might attend corporate brainstorming sessions.

Given this common diversity of viewpoints and objectives, two common outcomes prevail. In the one case whichever faction has the internal political clout sees its views prevail. In the other case debate produces a potpourri of objectives. In neither case is an intelligent focus of direction to be found. Only when the tradeoffs between conflicting constraints and objectives can be measured and compromised is a near-optimal strategy likely to evolve. Market conditions, focus of direction, finance possibilities and limitations, production realities, and cost conditions all must be weighed and balanced against one another to produce realistic plans.

Here, in the authors' experience, is the crux of the problem— the need for a realistic basis for planning. We have often encountered plans being drawn for equipment purchases that were unnecessary or unwise. We have seen products developed that could not be sold at a profit, acquisitions purchased that were unmanageable, and market expansions considered where no markets existed. The common thread between these strategic failures is a lack of a factual basis for decision making.

Some of these failures can be blamed directly on the C.E.O. of the organization. It is his role to promulgate and enforce the corporate philosophy. It is especially a failure of leadership when a firm adopts a mixture of objectives, selected to placate each department. More commonly the identifying characteristic of failed strategic planning is that the process flows solely from the top down.

TOP-DOWN PLANNING

Planning at the highest level of an organization is often an exercise in futility, a pro forma activity designed to make the chief executive feel good. Without a firm basis of technical information and advice from the operational levels of the business (bottom-up planning data), the upper level (top-down) planning process will almost surely lead to failure. There are two reasons for the failure of top down planning: (1) the lack of operational awareness already mentioned on the part of many upper level employees; and (2) the structure of the planning exercise itself. Consider a characteristic example of drawing up corporate plans by developing objectives and setting goals that seem to address those objectives.

The first step is determining just what the company's roles, missions, and objectives are or should be. Unfortunately, when these questions are asked of managers or workers the responses usually show an incongruous mix of normative and solipsistic attitudes. One person might say that the company should serve its customers in a way that justifies its usefulness to society. Another person might say that the company should do whatever it is doing right now.

Since the first round of gathering goals, roles, and missions is often so confusing, additional approaches are likely to be attempted. Some variation on obtaining group consensus is often attempted wherein managers are convened and asked to suggest a variety of objectives and then narrow the list down to the most important.

After a laundry list of broad objectives has been chosen as clearly acceptable, the next difficulty is translating corporate objectives into departmental and then individual objectives. The final product of the exercise is the assignment of explicit, measurable performance goals to everyone in such a way that the cumulative satisfaction of objectives will produce the desired corporate results.

Nominally, it appears to be a good idea to assign everyone a well-defined role in the company's operations. In fact, such an assignment should be made by top management as a natural part of its function. However, this is often ineffective, and can be counterproductive. The reason is that the goals set are in many cases not founded on fact. If the plan is built strictly from the top down, its goals are probably visionary but unachievable. Top-down planning will always be an intellectually bankrupt procedure unless it is built upon concrete bottom-up analysis

The two broad areas in which top-down planning fails are at the staff level and at the operational level. Few companies employ a rational system of staff evaluation. Little is known in general about the number of staff employees required for the company to perform at any given level of aggregate activity, and this ignorance makes it difficult to assign goals and objectives at the staff level. During recent recessions it became apparent to many companies, especially foundering firms, such as those involved in steel, autos, and petroleum products that their staffs were too large. This is a difficult problem to correct, since no one enjoys firing people, but many afflicted firms have reduced their white collar staffing levels. Yet, even now few formal systems attempt to employ objective evaluation criteria for white collar productivity.

At the operational level of the firm, top-down planning imposes goals that often make little sense. The authors have encountered goals such as ''limit late orders to major customers to less than one week'' and ''cut costs by 10%.'' Which customers are major? How do you define being a week late? The answers to these questions were not available from upper management. Neither was there an answer to: ''Cut which costs? What are the costs now?''

The difficulty in each case was that laudable broad objectives had been translated into baseless goals. The objective, keep our high-volume customers happy, was made into a nebulous operational requirement. The objective, cut overall costs, was transformed into a virtually meaningless dictate to the plant manager. However, these problems can be avoided.

Management must have the understanding and the data in hand at planning time that will facilitate implementable, factually based decisions. To do so will require bottom-up analysis—studies of what actually occurs and what implications this reality suggests for the future. At the operational level of analysis an excellent choice for a coordinator would be a strategically aware I.E. with line management experience. This is a person who knows what information to seek out and why it is needed.

Various experts describe the problem of top-down planning in different ways. W. Edwards Deming [9], the statistician revered by the Japanese as a guru of quality control, says that 85 percent of the problems in American industry are the fault of management, and he further says that the reason for management's failure is that the managers do not understand the jobs they are supposed to oversee. It is just this lack of understanding that is manifested as the goals and objectives of poorly conceived top-down strategic plans. However, it may not be necessary for top managers to understand the jobs beneath them in detail if they will simply delegate the task of deriving a factual planning basis to someone who does understand, someone like a strategically aware I.E.

The problem of understanding and evaluating staff and upper management performance is more difficult. There is, for example, no effective comprehensive productivity measurement system that measures each of the job categories: hourly workers, line management, staff workers, staff management, and upper management. Of these categories, upper management may be the most important to monitor because the direction of the firm is selected there.

SHORT-TERM MEASURES OF EFFECTIVENESS

Even if there is no genuine measure of upper management effectiveness, ad hoc measures spring up in its absence. Unfortunately for American industry, the usual measure is quarterly ROI, or at its most extensive range, annual ROI. A manager obsessed with this year's balance sheet and profit and loss statement is not one to think too far ahead. Not only will he ignore whatever corporate objectives the planning exercise has produced, he will give little attention to any kind of long-term strategic planning.

The attitude of the boss is passed on directly and indirectly to everyone else. Thus, the sequence: annual ROI as a measure of the boss's performance leads to short-term objectives on the part of the boss leads to short-term objectives on the part of everyone in the organization leads to an organization with no real objectives and no rational approach to strategy at all. This type of organization is usually characterized as being "myopic," and the term "myopic planning" is used to characterize this attitude toward planning. A firm that is habitually myopic tends not to invest capital in needed new equipment or projects because they increase liabilities or limit opportunities to distribute dividends that raise the market value of company stock. In short, the ROI criterion for managerial evaluation can destroy a firm's ability to formulate long-term plans.

WHAT MUST BE DONE

There are companies that do try to take the long view and that do profit from intelligent strategic planning, but to do this they must avoid the errors, cited above, that doom planning efforts at their inception. Effective long-term and short-term strategies start with intelligent objectives and these objectives can not be nebulous. A very good objective is: "Assure ourselves of adequate long-term profitability." The immediate question this objective inspires is: "How is the objective to be achieved?" That is where long-term planning comes in. Topics such as focus, finance, sales, and operations must be addressed specifically in the context of the company's major aim of profitability. Whether to set a numerical goal for profitability must also be determined.

Giving thought to long-term strategies will cause the intelligent manager to ask for information on costs, markets, facilities, and personnel. Only those long-term plans based upon physical realities have much likelihood of success. Basing plans on realities is the bottom-up approach that effective planning requires.

It is very easy to determine that a company has no logical direction, no plan based upon bottom-up analysis. One need only study the company's budgeting process. If the budget is not based upon planned operating levels and standard costs it will contain line items that arbitrarily increase from year to year. The more myopic the management, the more inscrutable the relation between funds budgeted and the purpose for their expenditure. This fact has inspired the advocacy of zero-based budgeting (ZBB) as a panacea for forcing rationality into budgeting, but ZBB applied to a rudderless company produces budgets as baseless as those produced without ZBB.

The extreme case of a company with no plan is that with no budget at all; it lives from hand to mouth. An unplanned, unbudgeted company inevitably incurs short-term debt at a rapid rate because it is constantly short of cash. Only when the C.E.O. understands a rational direction for company operations can a meaningful budget be derived.

An intelligent manager does not need to understand every detail of the corporate operation, he needs only to know enough to understand how things fit together, but he needs personnel who can create plans that allow for successful implementation at the operational level. Such personnel are exemplified by industrial engineers, especially those who understand corporate strategic planning.

THE ROLE OF INDUSTRIAL ENGINEERING IN STRATEGIC PLANNING

An objective analysis of the needs of bottom-up planning make it clear that individuals must be involved who can grasp why certain data are needed and how to implement plans. The industrial engineer, educated to the concepts of strategic planning, definitely satisfies the job requirements.

For example, the question might arise whether to expand production capacity for a given product line at a plant. One question to be answered is whether there is sufficient consumer demand to absorb the additional output. If this question were answered affirmatively, then before the expansion could be considered, two other questions would have to be answered. These are: can the plant run at the rate implied, and would the expansion pay for itself.

The way an industrial engineer would ordinarily investigate these two questions would be to conduct a feasibility study. He would study the plant in question and find out how it runs without the expansion. He would find out the average flow rates of machines, their capacities and those of storage locations. Then he would determine how the expansion should be laid out. He would then determine how the new equipment and the increased plant output would interact with the existing facilities. Then he would accumulate the costs of the various changes that would be required to effect the expansion, including any possible adverse effects on existing product output at the site.

The I.E. would then be prepared to perform a quantitative analysis that would cast various scenarios into physical and economic terms that could form the basis for intelligent decision making, but he could only do this if he had been informed of all the considerations that were to underlie the eventual decisons. Clearly, he would have had to be so informed prior to his data gathering effort were he to do an effective job.

Finally, the I.E. would produce a report explaining the probable consequences of available alternatives under the various scenarios. Scenarios might include higher or lower product demand than had been suggested, other ways to increase production, possible levels of capital costs, and different possible pricing levels. The I.E. would also explain what he had done, how he had done it, and what data were used in his analyses.

Given a logical study of the question of whether the plant expansion should be planned, a sound, bottom-up basis for a rational decision would be available to upper management. The elements required to achieve this happy state of affairs would be a willingness on the part of management to obtain a study of the plan, an I.E. who could comprehend what was needed, and an information-sharing network in the company that would permit data to be acquired and objectives to be understood.

Two of the three elements appear to be in the hands of upper management, i.e., the choice of who should be responsible for developing understandable policies and the responsibility for encouraging utilitarian flows of information. However, the educated I.E. must prepare himself and insert himself into opportunistic situations. An aggressive, persuasive I.E. should be able to educate management over time to the value of the work he is able to perform.

In some cases, even after decisions have been made, nothing happens. No action is taken by those in command. In one organization a reallocation of floor space that management had pronounced to be necessary to productivity improvement remained in limbo for over two years, an example of lack of follow-through. This is the single element of strategic planning that no one can affect but top management. If a corporate C.E.O. decides to pigeonhole a plan, or if he cannot make himself do the required work, the plan will die.

How can lack of follow-through be prevented? If the C.E.O. alone must initiate the action, little can be done, so responsibilities for implementing plans should not be left to the highest level only. In fact, the responsibilities should be placed as close to the action as possible. If an I.E. has been actively involved in preparing strategic plans, he would be the ideal individual to monitor the plans. He would surely be the proper person to monitor costs and the scheduling and methods aspects of plans, but he would also be capable of broader observation and supervision of projects.

One source of the failure of plans is the lack of follow-through at all decision levels after the decisions are made. This problem is made possible by the lack of measures of effectiveness that plagues most organizations. All personnel and all projects should be assigned criteria against which actual performance can be measured against projected performance. This includes the C.E.O. and all staff managers. All financial officers, accountants, salesmen, foremen, secretaries, staff employees, and vice presidents should be assigned specific measures of their performance. The I.E. should set an example by setting standards for his own activities. The I.E. department itself should prepare a set of goals for its work. In addition, if an I.E. helps prepare a strategic plan, he should also draft measures of effectiveness for that strategy so it can be evaluated over time.

Another way to encourage adequate follow-through is to prepare a timetable for achieving goals that includes observable milestones as checkpoints. Visibility of responsibility for action can be enhanced by making public who is responsible for each scheduled milestone. This strategy attempts to enlist peer pressure as a stimulus toward improved performance. An I.E. would be the ideal person to prepare these schedules and make responsibility assignments.

SUMMARY

The reason that strategic planning is done so poorly is that it is done top-down with little comprehension of the technical detail, costs, and operational repercussions representative of those policies under consideration. Active I.E. involvement in the planning process can lend to it the organized pursuit of the needed bottom-up analysis. However, I.E. involvement may not naturally evolve and may require persistent efforts by alert I.E.'s to demonstrate their value to planning and implementation activities.

PART TWO

THE POTENTIAL ROLE OF INDUSTRIAL ENGINEERS IN LONG-TERM PLANNING

Industrial engineers already play a role in long-term planning, although that role is not as significant as it should be. In this book's second part, the categories of long-term decisions, briefly discussed in Part One, will be examined in more detail to expose additional strategic ramifications. Examples of specific activities that bear upon strategic planning and that industrial engineers must understand to become more active participants are presented throughout.

A major theme that weaves its way through the remainder of this book is that of information systems. The success of all planning and implementation depends upon accurate and timely information. Examples will demonstrate this fact. The viewpoint of the following chapters is one of practicality. While vision and ingenuity are requisite to strategic planning, they would be useless without the guidance of practicality.

Chapter Four deals with the analysis of facilities decisions, stressing the role of capacity analysis in this category of strategic planning. Chapter Five discusses the more intricate topic of focusing strategies, and Chapter Six examines the unique strategic decisions that lead to acquisition and expansion. Each of these chapters expands on an important role of facilities in strategic planning.

4
FACILITIES DECISIONS

The adequacy of existing facilities and the need for changes in them are two complicated influences on strategic planning. Given a specific set of corporate objectives, facilities must be compatible with these objectives if the strategic plan is to be effective. It makes no sense to create a plan that has unrealistic expectations of the manufacturing facility. The strategic plan itself must depend in part on accurate analyses of existing capacity, current capacity utilization, and capacity requirements imposed by future plans under consideration. If the company's grand strategy is not based on a realistic appraisal of existing capacities and future needs, the strategy can easily fail.

Realistic data must be used to refine strategic plans iteratively until the plans converge to an acceptable compromise between vision and practicality. The need for bottom-up plans to forecast facility needs accurately is especially important due to the dominance of facility investments in the assets of manufacturing firms. These assets in turn define the strategic limits of production flexibility for their useful lifetime.

CAPACITY ASSUMPTIONS

Unfortunately, people rarely question the workings of familiar surroundings and assume the world is the way it is because that is the way it is supposed to be. One version of this world view inclines managers to believe that production facilities operate as they do because that is the only way they can operate. The authors have frequently encountered this attitude while interiewing plant personnel. A manager adopting this simplistic outlook does not ordinarily realize that he has made an assumption and will proceed toward facility decisions with an unrecognized bias.

As a major part of a modern I.E.'s work, he must often serve as an educator to coworkers from other disciplines. He must make it possible for decision makers to see the company in ways that they could not without the force of the imaginative application of his experience and knowledge. The basic job of the

I.E. is to make the unthinkable appear to be understandable and inevitable. To play this role in facility planning, the I.E. must confront implicit facility assumptions directly and make them explicit.

It is a rare production facility that has adequate data on hand to make intelligent facility decisions. The reasons are: (1) the tendency of people to adhere to habitual thoughts and procedures; and (2) their common acceptance of how their facilities work. Procedures and equipment that people feel comfortable with rarely excite enough curiosity to lead to disciplined data collection about them. An instructive exercise to perform at a manufacturing plant is to locate operational personnel, point to some important piece of equipment, and ask, "What is that machine's capacity? How much does it cost to run it? How much would it cost to replace it? What is its average availability? How much are its annual expected maintenance costs? How much power does it draw? How much scrap does it produce as a percentage of output? What labor standard is allocated to its operation? What percent of units it produces are subject to rework? What percent are rejected by quality control inspection?'

The answers to these little questions are always interesting. They range from quesswork to exaggeration. The authors regularly ask these questions whenever we have the opportunity, and rarely do we encounter knowledgeable responses. Yet their answers are the basis of intelligent operating modes, which in turn must form the basis for the process aspects of strategic planning, and, if he has been permitted to do his most valuable work, these answers are of constant concern to the Industrial Engineer.

A serious problem resulting from ignorance about facilities is that managers will make capital expenditures based upon merely fanciful notions of existing facilities. I.E.'s can help counteract this problem's effects by making their organizations conscious of real operational data, especially that pertaining to capacities. Corporate inertia ordinarily precludes upper management's assigning this task, so I.E.'s will probably have to instigate and conduct their capacity analysis efforts themselves.

CAPACITY ANALYSIS

What is capacity? What is maximum output? It is often reported in the newspapers that industrial facilities in the U.S. are running at some specified percentage of capacity. How is this capacity defined? The macroscopic percentage is probably obtained by comparing current production to the historical maximum. But should the comparison be made with respect to weight, or number of units, or dollar value? Each choice might yield a different number for a utilization rate. Again, what is capacity? Dollar output and volume output are not necessarily well correlated. Often a machine or plant is found to be more profitable if it produces a better mix of products, even though volume output may be smaller under the new mix.

In obtaining meaningful capacity numbers, an I.E. must obtain data for each product. Collection of only aggregate production information will not ascertain the features of each product, which can be expected to have unique requirements; one product may define the maximum number of units of production per hour while another gives the largest number of tons per hour. Capacities should be reported in at least two ways: as the maximum in units or tons that can be made and as the maximum profit that can be made from the facility. The twofold reporting allows interested parties to observe product tradeoffs rapidly.

The capacities for each product should be expressed as dollars of finished goods produced (cost of goods sold) and as anticipated dollars of profit (sale price less cost). One product will define maximum profit per unit of operating time while a different product will probably define the maximum flow rate. Products included in a capacity analysis should be those that will have the most impact on capacity planning; as usual Pareto's Law should guide the I.E. in the expenditure of his efforts.

The purpose of capacity analysis is to eventually converge upon accurate production goals. These goals must reflect not only capacity standards but also resource standards, such as labor hours per unit, materials per unit, machine time per unit, and energy use per unit. The I.E. performing capacity analysis must not allow himself to be distracted by the fact that standards already exist and are being used to budget and to track inventory and production costs. The existing standards may or may not be meaningful.

Some current standards may be useful numbers. A way to perform a rough check on existing production rates is to start with the measured maximum rate for the production of a given product, subtract the expected reject rate, subtract a rate representing expected rework, and then multiply the resulting reduced rate by the expected availability of the facilities used in making the product. The final result of this arithmetic should be a rough estimate of average production rate.

If the current standard (or average production in lieu of standards) is fairly close to this rough estimate, then the current rate is probably a good estimate of existing operational capacity. If the current rate differs significantly from the calculated estimate, the I.E. performing the capacity analysis must determine the cause of the divergence. Consider the following illustrative example.

Capacity Analysis

Product	Maximum Rate (unit/hr)	Maximum Rate Less Rejects	Reduced Rate Less Rework	Times Availability (90%)	Current Rate
A	15	12	11	9.9	10
B	20	19	17	15.3	11

In this comparison of two products made on the same machine, the calculated estimate of operational capacity is close to current rate for A but not for B. The current rate for A can be viewed as an acceptable number to use in planning production and studying facilities, at least until standards can be formally set. Product B arouses concern. It appears that either B's production is not well understood by the I.E. performing the study or that current practices in making B are improper.

These two simple examples demonstrate the two instances, one of being able to base capacity plans on existing operations and one of not being able to. One explanation for B's low output rate might be Parkinson's Law, which says, "Work expands so as to fill the time available for its completion." If time scheduled for completion of units of B were longer than necessary, the workers would simply take advantage of that fact and see to it that the schedule was never completed early. However, it could be that product B's production is somehow physically limited in a way yet to be revealed.

Once an accurate capacity analysis has been accomplished, intelligent facilities decisions can be addressed. For example, specific product mixes can be projected as being strategically desirable to market. Given market demand predictions and estimates of expected market shares the company can anticipate, facility capacities necessary to satisfy the market strategy can be estimated and the implied capital requirements can be projected.

The choice of precise focus of direction that a company should choose is reserved for a more detailed look in Chapter Five, but we can note here that market strategy coupled with profitability analyses may dictate a capacity mix that will lead to facility changes. These changes would be in addition to changes that natural growth or replacement dictate. Whatever the reason for facility needs, the accompanying decisions must deal with changes in physical capacities and their utilization. These decisions demand accurate factual bases if they are to contribute to productivity, efficiency, and productivity.

WHEN PLANS GO AWRY

Abnormal operations signify that the plans made for facilities have not succeeded. Several examples of abnormal operations, resulting from improper facility utilization, appear below. These situations should tell management that its facilities strategy is failing and give a few clues about how new plans should be drawn.

U.S. Steel decided to import slabs from British Steel for finishing at a U.S. plant that also had the capability of making slabs. This instance illustrates a facility that had lost its ability to perform one of its functions economically. The ensuing shift in strategy was to redefine the plant as a finishing facility instead of as a fully integrated plant.

A brick plant found itself shipping ground materials by rail from a sister plant some miles away. The reason was that the brick plant was capacity-bound, whereas the sister plant had surplus grinding capacity. The strategic decision was made to study whether some products could have their manufacture shifted from the capacity-short plant to the plant with slack capacity. The study by the I.E. staff showed that, not only could the change in facilities utilization be accomplished, it would improve the flow of operations at both sites. The strategy of shifting product lines was therefore adopted.

A steel mill found itself shipping slabs to a sister plant 1000 miles away to have them rolled into coils. The reason for this arcane maneuver was that rolling capacity was being fully utilized at the steel mill farming out the work, and the coils were among the few products it could produce that had an appreciable market. The cost of two-way shipping of metal cut profits markedly, and the plant struggled financially. The parent corporation adopted a strategy of selling the mill to its employees, having reached the conclusion that it could not be profitable if it were run the way the corporation was forced to run it.

In each case cited it is evident that facilities are not being operated in a sensible fashion. The obvious conclusion is that they are not being operated as part of a coherent plan. It is often difficult for an employee or manager to detect the fact that a procedure of his firm is illogical once it has been instituted. However, a good I.E. attuned to observing facilities utilization should be able to perceive and report on inappropriate procedures before they become irremediable.

The most common symptom of improper facility utilization is the presence of expediters on the shop floor. The immediate problem that requires expediting is poor scheduling, but the more fundamental problem is a lack of understanding of capacities and standards, which leads to a poorly conceived strategic plan.

As the example of the brick plant above shows, ill-utilized plants might be made to work properly if knowledgeable personnel were to investigate alternatives. In this setting an I.E.'s work is much like that of a detective, in this case searching for clues to better operations.

Just as shifts of operations from one plant to another can better distribute work loads, changes in capacity utilization within plants can permit more efficient production. As mentioned previously, capacities are contingent upon how facilities are operated, as are standards. One mix of operations will interact differently from another. It is, therefore, important for the I.E. involved in strategic planning to mentally experiment with the ways facilities can be used to eliminate sources of inefficiency, such as those found in the following example.

CAPACITY UTILIZATION

A bottling company made bottles for soft drinks. Its plants were dispersed to serve local markets more directly, and each plant produced a variety of bottles for its contiguous market area. A problem arose due to increases in demand for bottles. The company management determined that additional production of 3,000,000 per month was needed, but the capacity of a new plant to supply the extra bottles could not be less than 6,000,000 bottles per month. A smaller plant was ruled against as strategically unwise.

I.E.'s were asked to study existing operations to see whether additional capacity could be squeezed out of the old plants. The first step in their study was the determination of capacity standards. Realistic capacities were then determined from operating conditions to predict what production rates should be expected. As suggested above, estimates of average production rates were obtained by reducing the maximum capacity standards by subtracting down time allowances and average reject rates. The table below summarizes the data for the company's four plants.

Capacity Analysis

PLANT	MAXIMUM CAPACITY (bottles/mo.)	REDUCATION IN CAPACITY (bottles/mo.)	ESTIMATED CAPACITY (bottles/mo.)	AVERAGE PRODUCTION (bottles/ mo.)
A	7,200,000	2,500,000	4,700,000	4,200,000
B	8,800,000	3,500,000	5,300,000	5,500,000
C	6,600,000	3,000,000	3,600,000	3,800,000
D	6,600,000	4,000,000	2,600,000	2,600,000

This table shows that little capacity was left by the current product mix for new production. In fact, plant B was operating above its expected rate by extensive use of overtime. Nothing in the table seemed to refute the need for the new, expensive, underutilized new plant. However, the I.E.'s decided to study operations more closely.

The large differences between maximum and average production rates had aroused their curiosity. They discovered that the major loss of time was due to changeovers from one bottle type to another. The table below summarizes the I.E. findings.

Production Lost To Changeovers

Plant	NUMBER OF CHANGEOVERS PER MONTH	HOURS LOST DUE TO CHANGEOVERS	PRODUCTION LOST TO CHANGEOVERS (bottles/mo.)
A	128	107	1,280,000
B	241	216	2,600,000
C	170	143	1,700,000
D	280	223	2,600,000
		TOTAL	8,180,000

This table showed that over 8 million bottles per month of potential production capacity were being lost to changeovers. A study of the effects of grouping similar orders together and of redistributing production to partially specialize the plants was performed. The "best case" scenario of production improvement is summarized below.

PLANT	CAPACITY INCREASE BY CONSOLIDATING ORDERS (bottles/mo.)	CAPACITY INCREASE THROUGH PLANT SPECIALIZATION (bottles/mo.)
A	420,000	380,000
B	800,000	560,000
C	620,000	420,000
D	760,000	400,000
TOTALS	2,600,000	1,760,000

Altogether, an increase of 4,360,000 bottles per month could be squeezed from the existing plants, according to the I.E. analysis. In addition, it was found that the cost of changing production procedures would be considerably less than the costs of paying for a new plant. Management had only one difficulty with the plan.

Originally, the four plants had been ordered to make full lines of bottles and to react to local orders quickly. This policy had been part of the strategic plan of an earlier time, a plan that had stressed service. However, over the years demand had grown so fast it had created the need for new capacity, and management had to make a choice between continued profitability or a retention of the earlier service strategy.

Management pondered but finally rejected the new plant. They instituted the production plans suggested by the I.E.'s and tried to seek new ways to stress service that would not interfere with production. Times change; so do strategies.

This example emphasizes the need for bottom-up contributions to strategic planning, especially the kind of contribution that provides insights into alternatives that would otherwise remain unexamined. This is the kind of innovative thinking that I.E.'s are equipped to bring to the planning table.

INVESTMENT JUSTIFICATION

If the bottlers had decided to build a new plant, what criteria would they have used to justify the capital investment? Conformity to strategy would have been one criterion; building the plant would have preserved the long-standing commitment to local service. However, except in rare cases, an investment of such magnitude and repercussions would need to also be justified as financially viable.

Common financial standards emphasize return on investment (ROI) criteria as bases for investment decisions. Other companies insist upon immediate paybacks from investments; a company may not allow a capital expenditure that does not promise to repay itself within two years. This rule is myopic, and it would not logically fit into most long-term planning procedures. A more reasonable standard for accepting a proposal for investment is a long-term ROI criterion.

The two most common long-term ROI calculation methods are the discounted cash flow (DCFROI) method and the net present value (NPV) method. Both methods ordinarily discriminate between pairs of alternatives similarly, but each has its advocates and detractors.

NPV assumes some time value of money, expressed as an interest rate, and uses that rate to discount anticipated cash flows in the future back to present. The discounted cash flows are decreased by the cost of the investment, yielding the net present value projected from it. In the case of the new bottling plant, a useful life for the plant would have been assumed and NPV would have been calculated over that life. Other uses for the money required to build the plant could then be compared on a consistent basis by calculating their NPV's.

DCF assumes a useful life and future cash flows; then it requires that a discount rate be found that discounts cash flows so as to make the net present value zero. The interest rate so found is the "internal rate of return" for that investment and is usually compared to a minimum acceptable value to decide its adequacy.

Both methods of studying the long-term financial repercussions of capital investment are rational and both avoid myopic decisions, but both have limitations that must be understood. NPV guesses interest rates over a long period, and recent history has shown rates to have the potential for volatile change. Both methods project cash flows into the future, but both must project only marginal cash flows, those that are generated in addition to cash flows due to any other corporate activities and that are solely produced as a consequence of the new investment.

It is difficult to break cash flows in such detailed components because corporate units interact in ways that are often unpredictable. Understanding how to isolate specific cash flows requires a full knowledge of capacities, capacity utilization, and interactions of functions. This impies that financial analysis of capital alternatives should employ the advice of I.E.'s who do understand these facets of operations.

The citation of long-term ROI as a useful criterion for facilities decisions is not a suggestion that this method be used mindlessly to make decisions. Each facility decision should be considered holistically, weighing the various facets of the expected results of the each alternative. A facility modification must be operationally feasible, financially supportable, and within the framework of long-term strategy. The following example gives a good illustration of the diversity of viewpoints that must be brought to bear on facility decisions.

A TECHNOLOGICAL DECISION

A manufacturer of printed circuit boards has specialized in small-volume, customized circuit boards. The company uses two different methods of inserting electronic elements into the board: insertion by hand and by machine. The manual operation requires special fixtures in an efficiently designed work station. Automatic insertion by machine requires the preparation of a paper tape of instructions for each type of circuit board. The manual method takes 20 minutes to set up, while the automatic requires ninety minutes. Once in progress, manual insertion takes about 2 mintues per circuit board and automatic insertion takes 1 minute per board.

The company management decided that more insertion capacity was needed because that step had been identified as the bottleneck operation in circuit board manufacture. There was then one manual station and one automatic station for insertion. Management further decided that capacity would be increased by adding either manual or automatic capacity.

An economic analysis was performed for both operations of their capital and operating costs. Both methods of insertion satisfied the company's ROI criterion. A breakeven analysis was also done for each alternative to determine the range of run sizes to which they were well suited. Calculations consisted of:

$$\begin{aligned} \text{Total Run Time (manual)} &= \text{Total Run Time (automatic)} \\ 20 + 2x &= 90 + 1.0x \\ 1.0x &= 70 \\ x &= 70 \text{ circuit boards} \end{aligned}$$

The breakeven run size was 70 circuit boards, which indicated that the manual method was preferable for smaller run sizes while the automatic method was better for larger runs. The I.E.'s studying the capacity alternatives reported their results and made the recommendation that another automatic insertion machine be purchased. Their reasoning was that this choice would reduce labor costs and pave the way for eventually fully automating the plant.

Management rejected the I.E. recommendation. The reasoning that led to an expansion of manual insertion capacity was that the manual choice conformed to the company's strategy of selling to a specialty market. In such a market run sizes are typically small and changeovers frequent. To have invested capital heavily in equipment would have raised the company's breakeven point, forcing it into larger production runs. This, in turn, would have forced the company into direct competition with large, efficient, established circuit board manufacturers with whom it could not hope to compete. Management wisely decided to maintain its role in its traditional segment of the market.

SUMMARY

The company I.E.'s were not stupid in this case. They simply were not aware of the company strategy and its implications. They had been trained to believe in longer production runs and in the adoption of new technologies, but they had not been trained in strategic planning. If they had been aware of the broader picture they would have seen the larger arena in which the decision was played out.

At present few I.E.'s are aware of corporate strategies because management has not considered that I.E.'s need to know them. Neither have many I.E.'s interested themselves in the topic. The dangers inherent in their resulting ignorance are that management misses the opportunity to use the personnel with the best background for broad insights into corporate planning and that I.E.'s limit their own horizons by underutilizing their skills.

Lack of intelligent consideration of facility utilization, facility interaction, and the impact of facility decisions on strategic thrust is exhibited in companies that make myopic "here and now" decisions, those that slavishly follow ROI decision rules, and those that make absurd facility plans and violate their own best strategic interests. Strategically aware I.E.'s are the key to overcoming these errors.

5
FOCUS OF DIRECTION

A company without a thorough plan for its future has no plan and will suffer the consequences of aimless activity. A company, which follows a fully developed plan, will focus its efforts in those areas it understands, toward measurable goals that have been rationally developed. Intelligent focus of direction comes from the very top of the organization and begins with a broad statement of the company's mission.

Some choices of broad focus are:

We make a reliable, medium-price, medium-status automobile for the American middle-class consumer.

We sell rapid delivery of pastry products.

We sell chrome-plated automotive parts to automakers.

We publish high-quality trade paperbacks.

We serve the local needs for specialty steels.

These statements sound deceptively simple, but they may have required considerable soul-searching and deep thought.

The greatest temptation in business is to proliferate goods and services in a constant search for greater revenues. However, proliferation is destructive to well-focused operations, and, if it becomes a habit, losses and even financial collapse often ensue. This is why it is so important for a company, especially a manufacturing firm, to establish an intelligent statement expressing its focus of direction.

The broad statements of focus require ancillary statements of how the company is to view its role in the community, its stance toward customers, and how it intends to deal with its personnel. After all of these high-level objectives have been expressed, they must be tested against reality, and reality may force the statements to change.

It is common for a company's strategic plan to include "motherhood" statements, such as "We will increase our sales by 50 percent over the next three years." Unless this statement has been well-grounded in research, it will be merely so much noise. Increased sales must be founded on the ability to sell more product, the ability to make and distribute more product, the advisibility of doing so, and the profitability of the result. In short, unless an implementation plan for a strategy can be readily devised, the strategy is empty and invites the encroachment of unfocused operations.

An intelligent strategic plan must not only express a succinct philosophy of operation, it must be based on the most reliable bottom-up analyses of capabilities and opportunities obtainable. Capacity analysis, economic analysis, and market analysis are the cornerstones of corporate planning. Only when existing and planned operations are fully understood can broad philosophic strategies be translated into a functional focus of operations.

FOCUS

Given the broad objectives of the company, specific manufacturing, distribution, marketing, and support activities must be chosen as the functional mix of activities. The selected mix defines the operational interpretation of broad corporate goals. The test of the broad statement, in fact, is whether it is sufficiently precise to suggest a specific activity mix.

At a manufacturing firm its general statement of focus should define what it must do to be profitable for years to come. Thus, the statement must provide flexibility for adjustment to unanticipated changes in conditions but must prohibit spurious additions of unfocused activities to its activity mix. One should be able to consult the general strategy to determine the best choice of products and services, based on long-term considerations. A company with no long-term plan will chase any profit-making scheme without regard to how it may interact with everything else it does.

The ultimate test of a strategy is profitability, which consists of revenues less costs, and costs are the more difficult to understand. Historically, I.E.'s have been involved in cost control, so it seems logical that they should play a central part in profitability analysis, cost monitoring, and, ultimately, strategy. The questions about profitability that arise over time are (1) whether to change the existing product mix; and (2) what aggregate product mix would be optimal.

The search for the optimal mix of activities should be performed with faithful observance of all corporate objectives and long-term strategies, but it can be distorted by pursuit of short-term profits. Product mix questions sometimes lead to incremental product considerations that threaten focus, because its end result is the addition of yet another product or procedure to the current list.

Profitability analysis of new product, process, or service ventures is often difficult because costs are not available in a readily usable form. One resulting danger is that the I.E. studying costs will simply accept whatever costs are currently accepted without questioning their validity. A healthy skepticism should be brought to cost analysis.

It must be observed at this point that if costs are only poorly understood then the statements of corporate focus of activities will be merely fanciful guesses as to what should be done. There is no way to forecast profitability in the absence of accurate cost data. Therefore, formulation of corporate focus at a manufacturing firm must be based on accurate I.E. cost studies, and this implies that the I.E. must be conversant with corporate strategies so he will know how his study should proceed and why.

All studies of operating facilities must begin with the establishment of how they operate *now*. There also needs to be knowledge of the range of possible operations, i.e., the scope of potential flexibility. The determination of capacities and standards mentioned in the prior chapter is an excellent start toward the depiction of operations needed for a thorough cost analysis. Once costs have been established, market information can be used to derive projections of profitability for any given mix of corporate activities, but the first step in the entire analysis is the identification of the major corporate activities and their costs.

ACTIVITY ANALYSIS

Everything a company does costs money, but not everything it does brings in money, so it is especially important that management understands all contributions to costs. Otherwise, costs can grow incomprehensibly out of control. It is imperative that this not be allowed to happen and that costs be kept fairly predictable over time.

Such predictability requires an approach to costs that does not exist in many companies. It requires that the major activities of the firm be identified and that costs be associated with them. Many business consultants advocate an approach to cost identification that first separates the company into Strategic Business Units (SBU's), roughly dividing the company according to distinct products and services. However, such a division may be arbitrary and difficult to quantify. A better technique is the identification and quantification of natural activities.

Here the term "activity" is defined as any identifiable corporate function. Broad functional categories are accounting, marketing, and production of products, along with service functions such as the repair of previously sold units. Before an I.E. begins the activity analysis of a firm there is no definite number of activities he should expect to have identified at the end. Some activities turn out to be too minor to consider in depth; some activities naturally aggregate into compound activities; and others must be disaggregated into more manageable parts.

A good first source of activity identification is the company's accounting system. As a first cut, all ledger accounts can be treated as distinct activities. This classification will have to be refined as the study proceeds, but basing it upon accounting systems facilitates the tracking of expenditures, the sine qua non of cost analysis.

Eventually, the list of activities should be expected to include at least each final product, each consumer service, and each major internal support service, but a large number of steps may be required to reach this goal. For example, the Work in Process account for a product will initially be derived from the accounting list as an activity, but the WIP and Finished Goods inventories for each product should each be included as part of that product's combined activity costs. Although some commonly accepted accounting procedures may make this aggregation of costs difficult, it needs to be done. Once the activity classification is completed, it should be possible to list the costs and revenues associated with current levels of each activity.

One goal of identifying costs may be especially difficult; this is the allocating of costs to activities so that the total costs associated with the major activities account for almost all of total costs without a large amount of "unallocated costs" left over. This goal requires that the major portion of indirect and fixed costs be somehow allocated to activities.

COST ALLOCATION AND CAPACITY PLANNING

Accounting information may identify direct components of costs fairly well but may not deal with indirect requirements. At this point in nailing down descriptions of activities, pure industrial engineering of the most time-honored variety comes into play. Standards must be developed or verified.

Standards are resource requirements per each unit of operation of each activity. Direct labor, direct materials, energy, machine time, and any other important resource should have a reliable standard requirement for each unit of each activity. Pareto's Law should be kept in mind in order to avoid wasting time collecting insignificant standards.

Consider a foundry producing alloy ingots. Each type of ingot can be identified as an activity. Let us look at one such activity and its standards.

Ingot L-201

Labor	2 man-hours per ingot
Material	127 lbs. of alloy per ingot
Furnace time	1.3 hours per ingot
Setup time	1.6 hours per ingot
Finishing time	0.3 hours per ingot

These would be useful standards. If the number of such ingots produced were multiplied by the standards, the total direct resources required should be the result. Standards are based on the number of good units produced, so they automatically contain allowances for waste and scrap.

Given the standards for utilization of direct resources, the next step in activity description is the allocation of indirect resources and overhead. It is not usually difficult to allocate variable indirect resources to manufacturing; this is done by prorating costs to produced units according to their basis of variability, usually labor hours. The most common method for allocating manufacturing overhead is also according to labor hours. Thus, both variable and fixed indirect costs can be assigned to activities. For example:

Ingot L-201

Indirect Labor	0.3 man-hours/ingot
Manufacturing overhead	$1.44 per direct labor hour

Given the direct standards and the allocations, all costs of productive activities should be accounted for but general, administrative, and sales expenses. Activities representing these functions should also be defined and provided with cost standards. If new activities are to be considered, hypothetical standards and capacities will need to be estimated for them. The completion of activity identification should make it possible to obtain a fairly accurate estimate of total expenses for a given production schedule.

FULL ALLOCATION OF COSTS

Although the allocation of fixed costs to production units is an artificial device, it is not intended to be a public accounting procedure but one to help make rational decisions. The reason for costing out products in this way can be seen in a brief vignette.

The authors once were involved in the study of a printing company that printed a large number of special orders on several machines. The financiers of the company were concerned because the company was losing money at the rate of a million dollars per quarter. At a meeting consisting of one of the authors, company personnel, financing group personnel, and assistants, the company management was asked what profit margins their products had.

"Oh, accounting has that all worked out according to some complicated formula. Most products have a P/V of 25 or 30 percent."

This from a company losing a million a quarter. The problem that the P/V number causes is that it makes every company sound as though it is doing well. However, P/V (or contribution margin, as it is also called) is simply sales price less variable costs divided by sales price and expressed as a percentage. It represents how much margin above variable costs can be used toward defraying fixed costs. The implication is that a product with a high P/V must be a good product, but that notion is illusory. Consider a company making only one product.

Fixed costs	=	$100,000 per quarter
Variable costs	=	$2.00 per unit produced
Sales price	=	$8.00 per unit

The product sounds wonderful. It has P/V of 75%. However, only 10,000 units per quarter sell, so quarterly losses are $40,000. Now take the allocation approach to product costing. Prorating the fixed costs over the 10,000 units produces a total unit cost of $12, or $4 above revenues. This illustrates why only fully allocated costs should be used in studying the economic consequences of any given focus of activities.

ECONOMIC REASONING

As the above example demonstrates, many factors interact to dictate profitability. Some factors, such as product demand, constitute constraints that must be considered in drawing up realistic plans. Other possible constraints are imposed by capacity limits, resource limits, shortages of cash, long-term contracts, etc. These constraints interact with real costs and revenues to dictate the most economic activity mix.

To demonstrate the economic reasoning underlying the choice of an activity mix in light of constraints, let us look at a very simple example, which will also show why economic logic legislates against proliferation. The example is simplified to consider only five products as candidates for production and sale. The products are called A, B, C, D, and E.

The mathematics that underlie economic logic employ resource standards, cost standards, revenues, and capacities. These are exactly the data that capacity analysis and activity analysis plus market projections should have made available. These numbers for the example are shown below.

	Product A	Product B	Product C	Product D	Product E
Labor Hours	0.2/unit	0.3/unit	0.4/unit	0.25/unit	0.33/unit
Material Pounds	0.7/unit	1.0/unit	0.8/unit	0.9/unit	1.1/unit
Selling Price	$3.00	$4.50	$4.30	$3.80	$5.00

Labor Capacity—80 hours/week
Material Availability—400 lb./week
Labor Cost—$8.20/hour
Material Cost—$1.25/pound

With respect to the labor constraint the following maximum numbers of units can be made per week:

A—400
B—266
C—200
D—320
E—240

With respect to the material constraint the following are the maximum numbers of products to be made per week.

A—571
B—400
C—500
D—444
E—363

For each product the binding constraint is labor, so it defines the best product's production quantity in the optimal activity mix. To discover the best product in the mix, we will need the calculated profit per unit for each product, i.e., selling price less labor and material costs. These are:

A—$0.485/unit
B—$0.79/unit
C—$0.02/unit
D—$0.625/unit
E—$0.919/unit

The total profit that can be made by the application of all of the available labor hours to each individual product is shown below:

A—400($0.485) = $194.00 profit
B—266($0.79) = $210.14 profit
C—200($0.02) = $ 4.00 profit
D—320($0.625) = $200.00 profit
E—240($0.919) = $220.56 profit

The first product that should be made is E. In making 240 units of E, not all material is used, but, since labor is the binding constraint, this is the best solution, and only product E should be made.

The economic logic can be further illuminated by studying the value of the binding resource (labor hours) to each product.

A—$194/80 = $2.425/hour
B—$210.14/80 = $2.627/hour
C—$4.00/80 = $0.05/hour
D—$200.00/80 = $2.50/hour
E—$220.56/80 = $2.757/hour

This is a list of implicit values of the resource, labor hours, to each of the products, and these values dictate which is the best product for profitability when that resource is the most scarce.

It is not unusual at this point in a profitability analysis for someone to say, "But we can only sell 200 units/week of E." This new fact defines another constraint that the analysis must consider, and, since it turns out to be a binding constraint, there will be at least one more product in the optimal activity mix.

Reasoning through the new mix shows that 200 units of E will be selected first, based upon the value of labor hours to each product. This produces $183.80 in profits and leaves the following resources:

labor hours—14
material pounds—180

Remaining resources permit the following limits on each product to be produced:

Product	Limit Due to Labor	Limit Due to Material
A	70 units	257 units
B	46 units	180 units
C	35 units	225 units
D	56 units	200 units

Again labor is the binding constraint on each product. The profits resulting from making the permissible units of product are:

A = $33.95
B = $36.34
C = $ 0.70
D = $35.00

The second product chosen for the product mix will be B. Thus the amended product mix will be:

200 units of E

and

46 units of B

for a total profit of:

$$\$(183.80 + 36.34) = \$220.14$$

which is only slightly less than the profit in the absence of the constraint on E's demand.

Again note that two binding constraints exist, which in turn dictates a two-product mix. Also note that the second highest value per unit of labor hours was found for product B, which fact reinforces the reasonableness of selecting it second.

This abbreviated example demonstrates the fact that the number of binding constraints dictates the number of activities a company should partake in. A weaker form of this injunction says that the number of known constraints on company operations forms an upper limit for the number of activities in which a company should involve itself. The primary limitation of this dictum is that it only addresses support activities, such as staff functions, in a perfunctory way.

For example, sales of a product will be assumed to require advertising, so products will acquire cost standards per unit for advertising. The requirement of advertising is an assumed constraint, so advertising will be included in the optimal activity mix of most companies, even though its cost may be difficult to justify. On the other hand, unless R&D is also included as a cost allocated to the revenue-generating activities, it might well be excluded from the list of required activities.

If a company currently maintains 300 activities, including products, services, and support services but it can only cite 50 constraints on operations, it is involved in at least 250 superfluous activities, usually obsolete products. Whatever the source of the excess, serious thought must be given its reduction; mere habit does not justify it.

FULL RESOURCE ALLOCATION

As was briefly mentioned above, every function that is said to contribute to a company's primary (revenue-generating) activities must be allocated to them in such a way that an analysis of current activity levels indicates exactly where indirect and overhead activities are assigned. At the same time that support activities are allocated, the capacities of these functions should also be determined. This applies to all staff functions.

Organizations in which this kind of analysis has not been related to practice have been known to show disastrous consequences. For example, the federal government has discovered that it is capable of giving out grants and contracts far faster than it can keep track of them. A profit-seeking firm may owe its existence to avoiding such error.

All consumables in the company must be viewed as resources employed in activities. For example, the time available for staff members to perform work should be allocated somehow to productive activities so that total work time can be accounted for by activity levels and so planned activity levels will not require more than staff capacities can provide. The time standards developed can then be related to the accomplishment of line functions.

This particular resource allocation is often politically sensitive, so an I.E. trying to allocate resources to goods and services may be forced to simply prorate staff time over unit activity levels. The natural fear among staff personnel is the quite natural fear that their time will be found to be not well spent, which would suggest a cost-cutting opportunity through staff reduction.

Ideally, activity analysis allows accurate prediction of the costs and capacity utilization that would result from any given activity mix. When this accuracy of prediction does exist, any activity mix can be studied for its advisability under strategic criteria. The most common methods for making predictions about activity mixes and for choosing between them are linear programming (LP) and simulation.

MODELING METHODS

Simulation in this case usually consists of sets of equations and inequalities that must be simultaneously satisfied to depict the operation of the company under a given activity mix. The equalities and inequalities should be fairly accurate models of corporate relationships with respect to sales, production, costs, resource requirements, and capacity utilization. A simulation tool of growing popularity is spreadsheet software now readily available for microcomputers.

LP is similar in many ways to simulation in that it also requires that the company operations be modeled as equations and inequalities. The LP model can be so tightly constrained that it acts very much like a simulation model, but it ordinarily seeks an optimal activity mix subject to modeled constraints. A count of the equalities and inequalities in an LP model gives an upper bound on the number of activities that economic logic would select for the optimal mix. LP models are more commonly run on mainframes than on micros.

In either type of model the initially gathered data should be calibrated by studying benchmark cases. The best benchmarks to study are recent quarters of operation. The actual levels of profit and capacity utilizations can be estimated for these quarters as though they had not yet occurred, and the estimates can then be compared with recorded values. Major modeling errors can be caught at this point in the study and corrected.

After corrections are made the ensuing estimates of the benchmark cases should be so close to what was actually observed as to inspire considerable confidence in the numbers used. Once trustworthy cost and capacity allocations for the company's major activities are in hand, it should be possible not only to study specifically suggested mixes of activities but to develop new ones.

LP is a powerful technique for studying the effect of changes in policies and operating conditions. It is also the simplest and most reliable of the common optimizing procedures. If one is to generate a proposed optimal activity mix, this will probably be done via an LP model. The solution procedure for an LP model imitates the economic logic of activity selection discussed earlier. The drawbacks of LP are (1) it requires special-purpose software and a knowledgeable modeler; and (2) many managers do not understand LP results.

Simulation, on the other hand, can often be performed on business software in a simplified form. By its nature it is easier for managers to understand. However, simulation cannot ordinarily produce optimal results; it merely shows the consequences of a specific acivity mix. If there are only a small number of possible mixes, all can be studied and the best chosen, but the number of possibilities is usually large, so only a ''good'' mix can be found by simulation.

Product mix analysis is commonly performed at manufacturing firms, but aggregate activity analysis is not. Yet for strategic planning purposes all major activities, not just end products, must be modeled.

ACTIVITY MIX VS. PRODUCT MIX

This requirement of complete activity analysis requires special attention because the error of studying only end products is often committed. If only products, defined as units of production, are studied, then only variable costs will be developed and no understanding of other costs will be attained. This is the same error as that made by the printing company manager who cited his large P/V's.

The difference in emphasis between activity mix analysis and product mix analysis is very important for the I.E. to note. Activity mix analysis allows for the study of the company in light of its overall strategy. Product mix analysis allows for the study of a portion of the company's activities, mostly those pertaining to its manufacturing facilities. While it is not inadvised to study manufacturing alone, such a limited review should have a minimal role in strategic planning, which by its nature must be holistic.

Activity analysis must study the company's long-term plans. If a new product is being introduced as a major part of the company's strategic plan, its production will be necessary to the short-term plan even though it may adversely affect short-term profits. Activity analysis should be extended out over a reasonably long planning horizon, at least as long as that of a typical product life cycle, to study the long-term profitability, which product development seeks to enhance.

A strategic plan for activities may be fairly broad, indicating emphasis on a set of activity categories instead of emphasis on a detailed listing of specific activities. In this case the use of suboptimizing techniques such as product mix analysis may be entirely appropriate, and the broad policy guidelines can be treated as constraints on the product mix over a short planning horizon.

For example, the authors once helped a company plan its average product mix within the following guidelines:

Reserve 25% of capacity for arbitrary use
Make at least a given tonnage of products in category A
Maximize machine utilization

Given these guidelines we performed a standard LP product mix analysis which produced a listing of the optimal product mix. The analysis also showed that the capacity utilization criterion was superfluous due to the overriding effect of production bottlenecks.

In this example the company had chosen both production and marketing strategies as bases for its guidelines, but the results of the product mix analysis caused changes in both policies due to the new insights it provided. A broader activity mix analysis can be expected to create even wider enlightenment and more significant policy revisions.

The goal of comprehensive activity analysis is to provide the basis for choosing intelligent corporate directions. This basis can be ignored by unwise managers, and its value can be subverted by rapid changes in technology or markets, but it is an invaluable tool for strategizing. Another important aid is the analysis of capital productivity.

CAPITAL PRODUCTIVITY

When upper management discusses productivity, the specific topic under discussion is likely to be hourly worker productivity. There are two main reasons for this managerial view of productivity. First, American industry at one time was strongly labor intensive. Second, management is not usually inclined to ponder white collar productivity. There is a third measure of productivity that is also rarely discussed—capital productivity.

Recall that in the study of experience curves the point of interest was the decline of cost per unit of production as production accumulated. This number is the total cost per unit. Total cost can be expected to have the following components:

salaries
operating costs
materials
capital-related costs

Of these components capital-related costs have become steadily more important to manufacturing firms. The behavior of this cost component can be studied separately to produce measures of capital productivity.

This is a topic not readily broached by upper management because capital investments are clearly the responsibility of the manager who approved them. If they turn out to be unproductive investments, that manager would prefer that the issue not be raised. But if intelligent capital investment plans are to be made, capital productivity must be regularly evaluated. Poor investment choices can lock a company into an ill-advised focus of direction for many years.

One measure of capital productivity is ROI, but there are numerous definitions of ROI. One definition is:

$$\text{ROI} = \frac{\text{Profit}}{\text{Gross Assets}}$$

where profit could be either before or after tax and where gross assets are the original cost of capital items. However, this type of measure makes older investments appear better than recent investments of a comparable nature due to the tendency of more recent capital costs to escalate. An even more distorted measure of ROI is:

$$\text{ROI} = \frac{\text{Profit}}{\text{Net Assets}}$$

where net assets are original capital cost less accumulated depreciation.

A better measure of ROI is probably:

$$\text{ROI} = \frac{\text{Cash Flow}}{\text{Gross Assets}}$$

where cash flow is after tax profit plus depreciation. This measure allows the larger tax benefit of the later investment to be reflected and stresses the importance of cash flow, which is the true amount contributed toward working capital by company operations each year.

A more sophisticated measure of capital productivity is discounted cash flow ROI, the measure used to justify investments that was discussed in Chapter Four. The experience curve itself provides a measure of capital productivity. In a company that produces a large number of commodity items, that company should expect its own constant dollar costs per unit to decline, much as is expected for an entire industry. One reason the effect described by the experience curve exists is that more efficient technologies replace less efficient ones, but this replacement requires capital investment. If the investment does in fact lead to costs declining along the experience curve, then the component of costs related to capital cost (the capital recovery factor) can be said to be justified.

Many measures of capital productivity can be suggested, but the best criteria would allow for long-term review and planning. As has been suggested earlier, capital productivity cannot be measured without considering the effect of the investment on strategic focus. However, if investments are selected for their effect on overall ROI within the context of the company's grand strategy, poor investments should be few.

SUMMARY

This chapter has dealt primarily with the problem of choosing a rational, bottom-up, focused mix of activities given a relatively static set of facilities. This has been seen to be a natural topic for I.E. involvement due to the importance of factual operational and cost data to the selection of strategic focus. I.E.'s can further enhance their own value to the firm in this role by learning more about the kinds of strategy they will be involved in helping choose.

Other strategic questions that frequently arise are whether to acquire additional capacity by acquisition or construction and whether to branch out into entirely new products or services through acquisition or expansion. These decisions are special kinds of facility and focus decisions, but sufficiently dissimilar to the usual such decisions to warrant a separate chapter.

6
ACQUISITION AND EXPANSION

The preceding two chapters have emphasized static conditions, the analysis of existing operations, and the use of the results to direct the strategic planning process. This chapter is closely related to those two chapters, but it deals with dynamic aspects of planning, the rationale and timing of expansion and acquisition.

There is not a clear distinction between these two modes of change. An expansion of operations can take a company as far afield as an acquisition of an unrelated enterprise. On the other hand, acquisition of a competitor can represent little more change than expanding plant capacity. However, for clarity let us define the two terms:

Acquisition —growth through the purchase of or merger with an existing operating facility or firm with the intention of continuing its operation.

Expansion —growth by increasing the capacity or range of activities of existing facilities by building or acquiring additional facilities.

As with all definitions, one might interject an array of ''ifs and buts'' to cover special cases, but the thrust of the two ideas is clear. Acquisition implies the symbiosis of two organizations; expansion applies to the single organism. The fundamental question is how should growth by either mode fit into strategic plans? Do the plans naturally contain the flexibility for dynamic implementation or does growth suggest a revision of existing plans?

GROWTH IN STRATEGIC PLANS

Growth is a dominant consideration in strategic planning. The danger lies in unplanned, inadvertent growth. Growth connotes risk, since capital must be expended with no guarantee of reasonable return, while total lack of growth suggests timidity of management that does not augur well for the firm's health.

Naturally occurring conditions call for growth. Capacity analysis may show that profitable, growing demand for products cannot be satisfied with existing capacity; marketing plans that require penetration into new areas may similarly strain existing capacity. In either case the firm should consider whether expansion or acquisition might be advisable as a response to the challenge.

Company R&D might invent a new product that shows considerable market potential but which cannot be made with existing equipment or in existing facilities of the company. This event also calls for a rational analysis of expansion or acquisition.

A company in a position to do so may consider acquiring a company in a business complementary to its own, or it may consider branching out into such activities from scratch. Again the advisability of acquisition or expansion should be studied at the strategic level, based upon solid bottom-up information.

Successes resulting from expansion policies are legion, as are failures. In some rare cases expansion into activities far outside a company's expertise is rewarded, more often not. High tech products are now inspiring numerous expansions, acquisitions, and mergers. Some mergers will work to the advantage of both parties, but many will not. The reason may sound insignificant, but diverse companies, even in the same product line, may do things too incompatibly to mesh well together.

Expansions into new market areas or new product lines display a similar range of successes and failures as acquisitions. However, failures are not unexpected in situations that incur risk, so the mere possibility of risk should not be daunting. Companies that never take a risk also fail through inaction.

Successes have a chance to occur if the right questions have been asked and answered prior to the commitment to growth. Questions include:

How do we know the new market or product will repay our investment with growth?

What are the likelihoods of failure and success?

What are the alternatives to growing, and what are their likely consequences?

Instead of acquiring or expanding could we simply farm some work out to another company?

What are the results of market analyses?

What will the effect of the changes be on other company operations?

What will happen to overall profitability, ROI, cash flow, debt-equity ratio, and retained earnings?

If we acquire another firm, how will the two mesh to create a cooperative whole?

Do we have the management, staff, line personnel, and expertise required?

These are only a few of the questions to be asked, and they cannot be sloughed off as insignificant. All these questions should require the involvement of operational personnel who understand the interrelations between the operating units of the company and the overall strategic plan, in short, I.E.'s.

ECONOMIC, OPERATIONAL FEASIBILITY STUDIES

At the outset of any expansion two questions must be resolved: (1) Can it be done?, (2) Will it be profitable? An I.E. naturally approaches these questions through an economic, operational feasibility study. At minimum the background data needed is that produced by the capacity and activity analyses described in the preceding two chapters.

The average and maximum capacities that describe current operations and the standard resource utilizations and costs associated with existing activities should be readily available from the prior studies. Thus, the effects of expansion can be examined from a factual basis.

Consider a foundry contemplating whether to add a second electric furnace to enable it to branch out into a slightly different market for castings. It must estimate the cost of the addition, its output as measured in revenues, its demands on the finishing department's staff, its impact on the setup and pouring departments, and its interaction effect in reducing the output of existing products.

Estimates of the costs of additional activities can be produced by the economic phase of the feasibility study, but these costs are difficult to develop accurately. If fully allocated costs are available to describe existing activities, there will be a basis for similarly allocating all costs of the new activities; otherwise, without the allocation of fixed costs, incremental costs tend to become obscured as part of fixed overhead.

The feasibility study should report something like one of the following results:

The expansion is a good idea and should pay for itself according to accepted financial criteria.
The expansion cannot be fitted into the existing operation.
Market demand is insufficient to support the additional products.
The sand preparation room's capacity will be stretched by the added pouring requirements, but sufficient overtime can be tapped to handle that need.
The expansion is a borderline case: it may or may not be advisable.

There are many reasons an expansion may not mesh with existing operations, and these reasons can be entirely unexpected, which is why performing a bottom-up analysis, such as a feasibility study, is important. Consider the following example.

At a rolling mill the purchase of a continuous caster was under consideration. Studies of how the caster's output would augment the supply for metal coming in from current suppliers and how the added product would be absorbed by the market showed the caster to be feasible. Economic studies indicated the caster should pay for itself in a short time. Only when the production staff began to ask the caster manufacturer how the caster was to be operated was it discovered that the plant would not be capable of providing enough scrap to charge the caster's furnace. This totally unexpected technical problem made the caster purchase infeasible.

The rolling mill example demonstrates the fact that top-down expansion/acquisition decisions can easily create intractable operational problems. Only thorough economic, operational feasibility studies can expose the unexpected contradictions to strategic plans.

The steps in a feasibility study begin with gathering data. This is the first step without regard to whether the growth mode under consideration is an expansion or an acquisition. The data should include market projections, product price estimates, existing capacity, existing activity mix, resource standards, cost standards, estimated capacity of the new addition, resource standards for the new addition, cost standards for the new addition, capital cost of the new addition, and any additional data that suggest themselves.

Often it becomes apparent at this stage of analysis that resource requirements will exceed capacities if the new addition is put on stream. The product distribution network may be found to be overstretched, for example. In one instance a plant installed an expensive new machine's extra capacity to alleviate a bottleneck only to discover that the real bottleneck was the plant's material handling capacity, which fact showed the new machine to be superfluous. This problem would have revealed itself had a detailed feasibility study been conducted.

After data are in hand the new addition must be realistically modeled as a component of company operations. The new addition may cause a reduction in some previous activity levels, and, if so, this fact must be modeled. At this point of the study the adverse effects of proliferation show themselves; it is almost impossible to add activities without negatively impacting some other activities. To evaluate interaction, the entire operation, including expansion or acquisition, should be studied as described in the prior chapter, and then the new format's projected profits can be compared with the old.

Strategic decision making must consider many more factors than short-term loss of profit or temporary production problems, so an unfavorable short-term comparison of the new format with the old will not necessarily legislate against the new. For example, management may be quite sure that a major older product is entering the last stage of its product life, which fact demands that a new product be prepared promptly to take its place. This is the kind of forward thinking that motivates strategic planning. If an I.E. conducting a feasibility study is aware of this type of issue, he can more intelligently prepare the kind of study that will enhance the strategic planning process.

The above outline of how to study expansions or acquisitions will not vary from case to case, but significant differences exist between cases and warrant individual attention. Let us begin with the varieties of expansion.

MODES OF EXPANSION

It is a rare manufacturing firm that does not anticipate expansion of capacity as a predictable fact of life. Most firms expect to grow as a result of natural increases in product demand and as the need to diversify arises. The only points at issue are the timing and nature of the expansions.

Addition of machines to production lines and enlargement of storage facilities are two typical growth activities. These kinds of expansion can be classified as marginal expansions. When an entirely new product line requires a new facility or a new production line at an old facility for its manufacture, this expansion can be viewed as discrete. Discrete expansion will share some resources with the old facilities, but it differs from marginal growth in ways that suggest a unique set of strategic considerations. Since the two modes of expansion are distinct, they will be taken up separately.

MARGINAL EXPANSION

Marginal changes are difficult to evaluate economically because they incur costs and create revenues that are not easy to isolate. This is the result of interactions between activities. If marginal changes in operations are to be well understood, it is imperative that the total resource comsumption be allocated to activities.

It is not unusual for a marginal change to cause a resource limit to be reached unexpectedly. It could occur in the fabrication department, in accounting, or in distribution, or a production step that previously had slack capacity could suddenly become capacity-bound. Without accurate understanding of resource consumption by activities, these problems can occur without warning.

Even after careful preparation, predicting the effects of marginal changes can be difficult. A case in point is a radiator repair shop that decided to expand its services by offering to repair oilfield radiators used to cool drilling pump motors. The unanticipated aspect of this expansion was the size and weight of the radiators to be repaired. They tended to range above four feet in height and over a thousand pounds in weight.

The shop had only one heavy-duty winch and little floor space for dismantling the radiators, little bench space for recoring them, and no basin large enough for tank cleaning. In addition, the delivery trucks for the shop were not well suited to the pickup and delivery of the large units. It is doubtful that an objective inquiry into the hows and whats of the new service would have allowed it to proceed past the discussion stage, but, once entered into, it provided everyone at the shop with daily logistics problems.

How could a company with trained I.E. personnel justify making the kind of error the radiator shop made?

"Well, we didn't see any need to do some long-winded, time-consuming I.E. study" is a common response.

A company's ability to make strategic plans becomes restricted by facility decisions. Once a service or product is entered into, habit takes over, and the additional activity is difficult not only to abrogate but even to question. The best time to delete unproductive expansions is prior to their adoption.

Since unexpected complications are the danger of expansions, an I.E. studying a suggested one must try to bring possible problems to the surface so they can be discussed at the strategic level intelligently. It is not sufficient to wave a vaguely defined red flag. Consider the oil and gas exploration firm that was looking for additional revenue-producing activities.

A member of the Geologic Analysis division suggested that the company possessed the finest computerized geologic database and field analysis system of any oil and gas company in its region. He suggested starting an information service to other companies for a subscriber fee. Since the company already had the system in place, there would be no development cost, so the profits should be immediate. What red flags should the staff I.E. have raised? Some obvious questions he asked were:

How is the service to be promoted to attract subscribers?
How much can be charged that will make the service affordable to subscribers?
How much must be charged to justify the expenses of running the service? What ROI does this represent?
How much will the marginal expansion of geologic activities cost to perform?
Will this service detract from the promptness with which Geologic Analysis responds to our own needs?
Do we have the expertise necessary to manage such a program?

In this instance the questions were not able to be answered in such a way as to justify the expansion. The last question was the most critical, for it was discovered that not one executive in the firm had the slightest idea of how to begin such an activity. If it had not been for the I.E.'s little questions, the company might have entered into an expansion that was certain to produce unexpected results. Unexpected results have a way of creating a mad scramble to respond to them, and mad scrambles do not fit into intelligent strategic planning.

In summary, any marginal expansion must be justified on the bases of satisfying long-term strategic objectives and long-term profitability, but the profitability must be judged as the increase in profit over that obtainable without the expansion. Thus, if an expansion reduces resources to other activities sufficiently, it could turn out to be only slightly profitable or even loss-inducing. ROI must be calculated by using marginal cash flow as a return on the expansion's total cost, including costs required of other activities to accommodate the addition.

The fact that other activities must be considered in measuring marginal profit and ROI may seem trivial, but this point is lost on many managers. An I.E. is trained to expect interaction effects between activities; a manager not fully familiar with operations may not have acquired this predilection. The practical effect of management's lack of understanding of this point is that expansions are often measured by economic evaluations that treat the additions as independent of existing activities.

Many marginal expansions result from simple extrapolation of recent sales; a continuing climb in demand for a product will eventually strain any production facility. This expansion motivation incurs a lower risk than does a speculative expansion; entering new markets or opening new market areas are both more speculative than growth resulting from trend extrapolation.

There are two commonly accepted methods for weighing the risks of expansion. Since these methods apply to all kinds of expansions and acquisitions, it seems appropriate to digress briefly to present these risk evaluation techniques. They are the scenario and the expected value techniques.

SCENARIOS

Scenarios were obliquely referenced in the previous chapters. A scenario is simply the description of the full consequences of a set of assumed conditions. In a contemplated expansion of market volume or market area or a planned diversification, the likely scenarios to consider include "best case" and "worst case." Intermediate cases should also be studied.

A "case" is a situation that hypothetically results from specific values of the parameters defining a possible future. In market expansion of any sort, parameters include consumer demand, prices of products, and costs of delivered products. Other parameters include adjustments to the other activities affected by the change in procedures. A "best case" minimizes uncertainties and ill effects, such as lower-than-anticipated demand. A "worst case" on the other hand suggests the lowest anticipated return on investment. A rule of thumb might be to support any expansion that a worst-case scenario finds acceptable.

EXPECTED VALUES

The expected value approach is best applied when the list of possible situations is quite short, whereas scenarios are applicable even when a continuous range of possibilities is envisioned. For example, let us concoct a brief example of expected value analysis.

Say a company finds that exactly three cases are possible:

Case 1: Expansion will lose \$100,000.
Case 2: Expansion will gain \$10,000 profit.
Case 3: Expansion will profit by \$300,000.

The expected value derived from the three situations is the weighted average of profits. To obtain this average requires knowledge of the probabilities of each situation. Where these probabilities may be obtained is hard to say; in some cases past data can be used to prepare average rates of occurrence that seem to be reasonable probability estimates. Say in this case that we found (or guessed):

$$\begin{aligned} \Pr(\text{Case } 1) &= 0.2 \\ \Pr(\text{Case } 2) &= 0.1 \\ \Pr(\text{Case } 3) &= 0.7 \end{aligned}$$

then the expected value of profit would be:

$$\begin{aligned} E(\text{profit}) &= -0.2(\$100{,}000) + 0.1\ (\$10{,}000) + 0.7(\$300{,}000) \\ &= -\$20{,}000 + \$1{,}000 + \$210{,}000 \\ &= \$191{,}000 \end{aligned}$$

This expected value would have to be compared with the required investment to determine its desirability, but at a glance the numbers seem to suggest that the chance of making a large sum is far more imposing than that of losing money or of virtually breaking even.

Scenarios generally appeal to managers more than do expected values, probably because scenarios duplicate the way most managers mentally examine alternatives. Whichever procedure is used, analysis of risky alternatives is as readily applicable to one type of contemplated growth as another. It applies to acquisitions, marginal expansions, and discrete expansions, which are discussed next.

DISCRETE EXPANSION

This topic is perhaps more difficult to analyze for managers because they tend to see marginal and discrete expansions as dissimilar animals. In fact the two differ very little. Consider their like characteristics.

Both modes of expansion require more money, both for capital and operating expenses. Both aim at increasing revenues. Both involve risk. Both will require higher operating levels of some company activities, including those at the management level. Both marginal and discrete expansions share operating capacities with the original operating facilities. Both run the risk of unpredictable repercussions.

The main differences are that discrete expansions usually play for higher stakes than do marginal ones. Possible gains achievable from wholly new markets or innovative products are often enormous. Numerous products over the past three decades have demonstrated the potential of bright new ideas in product development and marketing. Success stories include Xerox, IBM, Lite Beer, and Japanese automakers.

Discrete expansion has unique features. It commonly inserts the company into a situation with which it has little familiarity. Expansion into either new markets or expansion into new facilities both tread the unknown.

When new market areas or new product markets are entered they must be preceded by market surveys. If the company does not have the in-house capability to perform them, there are competent firms, which do little else, that can be hired for the job. Simple market extrapolations or guesses will not suffice to justify discrete expansion.

The financial justification of a discrete expansion by NPV or DCFROI methods will require a projection of the useful life of the investment. This life is the shorter of the expected retirement period for the facilities and the expected life cycle period for the product motivating the expansion. In the engineering economy calculations that I.E.'s learn in school, a facility with remaining useful life at the end of a period under consideration is commonly assigned a salvage value, so, if the product life cycle is shorter than the facility life, the facility's remaining utility can be expressed as an positive salvage value during the financial calculations.

Even after the most careful market, operational, and financial analyses have pronounced an expansion to be strategically viable, it may still fail due to unforeseen events. This is the nature of risky activities; some failures are sure to be encountered. However, the most common sources of unforeseen problems are interactions between activities competing for common resources. One method that has been suggested for minimizing these unpredictable problems is the "plant within a plant" expansion [69].

Say a firm making cardboard containers decides to install a specialty production line that will make a type of carton distinct from those already being made at the existing facilities. A way to avoid unforeseen conflicts due to resource utilization conflicts is to introduce the new production line as an entity insulated from the rest of the plant; this is a version of the plant within the plant concept.

Isolation will require the new line's paper supplies to be ordered separately from those of the rest of the plant's orders. All other supplies, such as glue, will be similarly separated. No workers will be shared between the original plant and the new line; no storage, shipping, or operational space will be shared; heat sources will be separate. As far as is possible no management or staff personnel will be shared.

Since the plant within the plant will probably have a smaller level of production than the original plant, its staff structure is not likely to imitate that of the larger operation. For example, a single person might perform both the accounting and order entry/release functions for the new line. Some sharing of staff is unavoidable, however; sales should be fully coordinated according to the strategic plan for the new collection of activities, so the sales staff should be shared by the original and new products.

Even if the plant within a plant concept is not appealing or is inapplicable to a given expansion, the lesson remains that unforeseen conflicts between old and new activities must be held to a minimum in discrete expansions. This requires thorough advance study, the kind I.E.'s understand.

In summary, expansions must be prepared for in a rational way. They cannot be viewed as completely divorced from existing operations. A glaring example of the interference a new product can create is given by a company that manufactured gas space heaters. In a quest for diversification, the company decided to make kerosene space heaters. After a few months of marketing their revised product lines, the salesmen discovered that the company was competing with itself; people would come into a store looking for a gas space heater, and, seeing a trusted brand of the cheaper-to-operate kerosene heater, they would buy it instead. Gas heater sales plummeted while kerosene heater sales rose, but the total profit dropped because the kerosene heaters were not yet far enough down the experience curve to show low manufacturing costs.

Any increase in complexity of an organization, whether in management structure or in product lines leads to unexpected problems. This statement is equally true when specialized to either expansions or acquisitions. Thorough planning and analysis and intelligent strategic planning can avoid the most debilitating consequences in either case.

ACQUISITIONS

Acquisitions can be categorized as either marginal or discrete, but, given the preceding discussion of the distinctions between these two categories in the case of expansions, both can be treated simultaneously here. All warnings about resource sharing and possible adverse effects of interactions in the case of expansions also apply to acquisitions, but additional cautions must be cited.

There are two ways for a merger to occur: cooperatively and adversarially. If a merger is cooperative, it may consist of little more than a loose combination of independent parts unless the two companies are put under a single managerial structure. If both prior management structures are maintained virtually intact, the terms of the agreement between the new partners will dictate how the merger will work, and these terms vary widely.

If an acquisition is of the unfriendly variety, the acquired company is sure to be put under the thumb of the acquirer. The management staff of the purchased firm becomes subsidiary to that of the parent. Conflict will inevitably arise from all but the most cooperative mergers over chain of command, cost control, objectives, markets, and budgets.

Machiavelli [20,14] suggested that in an acquisition the parent company would either have to accept the limitations presented by the acquired staff or replace it. This advice still seems sound. The staff of a new acquisition cannot be expected to promptly adopt a new array of habits. Old habits will conflict with parent company culture; however, replacing the staff of a company whose expertise lies outside that of the parent may create even worse problems.

ACQUISITION CHANGEOVER

Staff friction is a sure source of difficulty following an acquisition and cannot be relegated to the role of "unanticipated" problems. Other difficulties may also arise. Yet there are good reasons for considering acquisitions. The question is how to make them work. Consider a cooperative merger first.

A merger may be motivated by complementary products, resources, or facilities that are felt to be beneficial to the objectives of both firms. A computer firm may merge with (or cooperatively acquire) a communications company for the purpose of rapidly entering the information systems business. Two dissimilar disciplines seem to be required for the targeted business, which disciplines would be difficult for either firm alone to develop. The merger is superficially attractive and reasonable, no guarantee that it will also be tractable.

Many seemingly reasonable complementary mergers have failed. Probably the basic cause for the failures has been the difficulty of coordinating staff and management activities. As soon as diverse units are brought to a common goal there must be a unity of purpose brought to all procedures that reflects a common strategic plan, which in turn must spring from a cohesive planning process. Unless the interface between the two parts has been thoroughly addressed prior to the merger, it may be too late to patch them together post-merger.

For example, in the case cited above of the newly formed information systems company, how should each half revise its activity mix? Since there is a de facto commitment to a new product line on the part of each partner, it is clear that some previously important activities will have to be cut back. The problem is actually that of ascertaining a two-firm optimal activity mix under their individual and common constraints. The uncertainties attendant to the formulation of this problem are formidable.

If two facilities within the same company have shared objectives, it is not too difficult to solve a two-plant optimal mix problem because data from the two parts are probably similar. Even data-reporting forms from the plants may be alike. However, two diverse companies just prior to a merger are unlikely to find themselves with these favorable conditions for merger analysis.

Assume that a pre-merger activity analysis is to be performed. Who will perform it? If an I.E. from the computer company suddenly appears upon the doorstep of the communication company prior to a merger agreement, will he be invited in and shown all their books by the accountants? Probably not. When then will it be possible to perform the analysis that predicts whether the merger will be profitable? After the merger.

When will managers get together and seriously sort out who has responsibility for what? After the merger. If the merger is preordained to fail, when will clues of this fact begin to surface? Too late. It must be said in the defense of cooperative mergers that there is a fair chance for most of them to succeed for two reasons: (1) indebtedness of neither party is increased by the merger; (2) both parties are committed to some common goals. Even if the merger will not work, it can be readily dissolved without an exorbitant loss of investment. Neither of these advantages occur in unfriendly takeovers.

AGGRESSIVE TAKEOVERS

Many recent takeovers of one company by another seem to be motivated by specious motives, questionable at best. A company short of cash may swap stock or may accept additional indebtedness simply to obtain the cash that a cash-rich company has on hand. This short-term objective does not translate into long-term strategy. Many other takeovers appear to be motivated by little more than corporate aggressiveness.

On the surface an acquisition may sound reasonable but may be little more than an exercise in territorial expansionism. The desire to have the company grow larger is not uncommon in the boardroom. Among a group of executives with this mindset, questioning the advisability of an acquisition will not meet favor. An I.E. who finds himself mired in a company motivated by an adolescent craving for bigger holdings had best begin his job relocation search. Rationality will not be heeded in such a setting.

A takeover can be logically motivated by economic advantages or by technical advantages. A chemical company acquiring an oil company is explained as guaranteeing chemical feedstocks to the acquirer. An auto manufacturer taking over an electrical company can be explained as hastening the development of economical electrical car motors. The takeover of one food conglomerate by another is explained as fleshing out certain product lines.

The only genuine explanation for an acquisition is that it improves the acquirer's ability to meet its strategic objectives. Whether the explanations in the cases cited above can be so interpreted is unknown and uncertain. In fact many takeovers have the character of mere paper activities; an acquiring firm buys another firm, much as it would a stock, deriving dividends from it and disposing of it when it has suitably appreciated.

A paper takeover does nothing to advance a strategic plan, since it is merely a mode of temporizing before deciding how to use investment funds. The argument can be made that every company goes through ups and downs and that, when the internal rate of return is temporarily down, it should invest for the time being in something other than itself. This argument is questionable; any company that finds it impossible to make a return on capital exceeding that of outside investments should concentrate on restoring competitiveness.

In a takeover for some purpose other than paper profits, not only is the activity mix of the merged units difficult to analyze beforehand, there are several technical decisions that must be made initially on scant evidence. One such decision is whether acquired operating units should be run as cost centers or profit centers.

COST CENTERS AND PROFIT CENTERS

Most companies operate as a mix of cost center units and profit center units. A cost center has a budget it must adhere to and a set of duties it must perform. A profit center operates somewhat independently, determining its own activity mix, setting its own prices, determining its own product strategies, and trying to enhance its profitability.

If a cost center acts as a supplier for a profit center within the same corporation, it will charge transfer prices for the items supplied at levels intended to recover the costs of the cost center. If a profit center sells items to a division of the company, it sets its prices as though it were selling outside the company.

In an acquisition, the parent company often changes the classifications of several corporate units from profit centers to cost centers. There are at least two reasons for these moves. First, the autonomy of profit centers may be frightening to the parent company's management. Second, the anticipated lower cost of parts supplied by a cost center was probably a major factor motivating the takeover.

The disadvantage of switching an operating unit from a profit center classification to that of a cost center is that the morale of its personnel will plummet and its creative zeal will evaporate. A compromise solution to reclassifying profit centers is to allow the divisions they supply to go into the open market for parts. If the acquired profit center had the parent company's divisions as customers before, it will be forced to keep its prices in line with outside competitors while maintaining its drive to gain an edge on them. Another advantage gained by not reclassifying a profit center is that this move reduces the disruption of the unit's previously devised strategic plan, and the sudden disruption of such plans at all levels is the overriding problem that acquisitions create.

CAVEATS

Acquisitions are seductively attractive. They appear to present a ready source of cash, or a low-cost supplier, or an instant mode of diversification, or a one-shot beautification of the balance sheet. The only valid justification for an acquisition is that it represents a more advantageous mode of satisfying strategic needs for growth than does expansion.

Acquisition is much more uncertain than expansion. None of the requisite capacity and activity analyses will be accessible for the company to be taken over, so there will be no way to perform profitability analyses or studies of activity mixes before the fact. The after-acquisition management structure is uncertain, as are the post-takeover strategies for the newly-formed conglomerate. The true financial position of the company to be taken will not be known. If the implicit diversification the enforced merger represents is too far from the acquiring company's skills, it cannot be expected to run the acquisition intelligently. The bottom line is that unless an acquisition offers outstanding benefits, it should not even be considered.

SUMMARY

Rational criteria tend to support expansion in existing or related product lines if profitability analyses and risk assessments indicate reasonable likelihood of adequate return on investment. Unless overwhelming advantages are presented by mergers or takeovers, they should be avoided as essentially unpredictable and usually unsuccessful. An I.E. involved in studying growth plans will probably have the opportunity to have a strong role in presenting the facts.

When a manager fantasizes about a new activity he has proposed, he may feel he is hitching his future to it and feels, therefore, inclined to compare it to one of the success stories. He may not visualize operational and financial realities quite so readily, and many companies abound with personnel who are loath to argue with their boss's illusions. An I.E. might reasonably ask, "Why should I be the one to puncture the balloon?"

The answer is, "Because that is your function. If you shirk it, it may not be done. You have to take responsibility for injecting rationality into strategic planning whenever it is missing." If the I.E. never finds rationality guiding policy in his company, perhaps that is not a good company to remain with.

The I.E.'s role is that of a provider of realistic plans and appraisals. If he consistently demonstrates that he fills this role, his input will not be taken amiss by intelligent managers. Further, if he shows cognizance of the strategic implications of his appraisals, he may gain the position in strategic planning that he logically should occupy.

7
RESEARCH AND DEVELOPMENT

A question that seems implicit in the phrase "research and development" is: Why are these words always used together? That is a good question; are they so closely related that there is no way to separate them? One is reminded of a chimpanzee in a zoo, lying on its back fruitlessly trying to peel the layers from a single leaf of lettuce. R&D, is it a single entity?

The answer is actually "No." The probable reason the two are lumped together is that their roles in a company's activities are similar. Both are commonly orphans. Few divisions or managers want responsibility for devising new ideas, nor does anyone want to explain that R&D money was spent but produced no tangible results. Few companies treat R&D as necessary and accountable functions, so their importance has diminished in American business over recent decades. Even large firms, notably those involved in transportation, energy, and primary metals spend small percentages of their revenues on R&D.

The lack of R&D investment in the American steel industry is about 0.7% of sales [68]. This fact is especially puzzling, since American steel has become steadily less competitive with imports. Lack of R&D commitment is not so surprising in small firms. R&D appears to many managers to fill no clear need in the organization, which fact makes them orphans. Also, managers in search of short-term profits may discover that R&D are easy to cut out of the quarterly budget. Thus is the life of orphans.

STRATEGIC PLANNING AND R&D

The existence of long-term planning suggests that changes occur in corporate behavior over time. Rational change requires rational justification, which in turn requires some effort to develop. That is, strategic planning requires some form of the organized investigation called research to exist. It may be that managers have compartmentalized research in their minds to consist only of arcane "pure" research unrelated to practical objectives and do not perceive operations research and market research as research activities. If this is the case, these managers need educating.

A company can approach the market with innovations of its own, or it can rely on being imitative of competitors. The latter strategy requires no research and little development, but it is a stagnant, noncompetitive ploy. Basic economic theory shows that late entries into markets can never catch up to leaders in cost control nor in profitability. This means that followers will always lag behind market leaders in cash flow per unit of product. The leaders get richer and are able to invest more in innovation; the followers drop ever further behind.

Following others involves very little strategy—whatever the leader does, the follower is forced to do. Just as a rebellious youth has his actions dictated to be the opposite of whatever he is rebelling against, a company that only imitates is dictated to by market leaders. It is not hard to see a large leader company driving a small follower out of business by tricking it into ruinous actions. A sadistic manager of a leader might do this just for fun.

R&D are a company's insurance policy against the pitfalls of mindless imitation. However, managerial support for R&D is not an accepted management fad at present, so America's real innovations are being concentrated largely in small, entrepreneurial firms whose new ideas are their life blood. An I.E. employed at a larger, static company will find himself very rarely involved with research and development, whereas he will be busy with R&D issues at an entrepreneurial firm.

Many larger concerns have recently begun to copy smaller firms' research models by establishing small units inside the corporation and entrusting them with specific projects; IBM has taken this tack [71]. This may be a good way for a large company to approach focused issues, but it may avoid the larger question of how to establish a continuing, effective R&D effort.

Research and development are topics that are rarely considered within the accepted charter of Industrial Engineering. However, the long-term effect of R&D makes it essential for the I.E. to be involved in this important phase of business activity.

RESEARCH CATEGORIES

Too many managers visualize research as something akin to the activities of mad scientists in monster movies, pure research with no practical goal. *Pure research* is a valid and necessary type of investigation, but its goals should always be well-directed. It is difficult to visualize a major manufacturer of computer chips not researching vigorously to discover fundamental breakthroughs in chip design and materials. Much of this work could be classed as pure research, but applications of the results are not difficult to imagine.

Many corporations devote a great deal of pure research to materials, looking for cheaper, stronger, or lighter components for their products. However, pure research cannot be as accurately targetted as development, and this is an aspect of pure research that managers dislike. Cancer research presents a good example of the necessity for pure research to be allowed freedom; many important pieces of knowledge about human cancer have resulted from general microbiological research into cellular knowledge. Cancer resesarch was not the prime motivator of this work. Simiarly, no one had small tv sets in mind when the transistor was discovered in the 1950's, but it seems clear now that a company dealing in electronics devices is well-advised to support general research in electronics.

Whether a company's research effort is classified as pure or applied does not mean that every piece of research done has to be readily placed under one banner or the other. But all research should have one characteristic in common—it should be directed by well-defined strategic objectives.

Applied research is more readily palatable to managers. It sounds like an activity trying to make money. Nominally, applied research is aimed at solving specific problems of manufacturing, but, as is the case with pure research, some of the activities may seem unrelated to current concerns. A rolling mill may find itself assigning metallurgists to the study of metal structure in order to learn how to roll metals that won't break under fabrication methods at its customers' plants. Is metal structure research pure or applied?

Instead of insisting upon a cut and dried scheme for sorting investigations into one of the two camps, it might be better to define research as the organized investigation into topics of future strategic importance. What topics might these be? They should include all topics of both long-term and short-term planning significance. At minimum, research and development topics should include new or revised products and services, methods and equipment, materials, operations and organization, human factors and incentives, markets and marketing, and quality assurance and monitoring.

Two problems restricting the employment and effectiveness of research and development are the lack of management commitment to them and the poor management procedures applied to them. R&D are not always perceived as necessities by managers of businesses because their importance to strategy is not understood. Correcting this lack of understanding is a challenge that I.E.'s must try to overcome. One way to attack both the lack of interest in R&D and its poor management is to put R&D efforts into a framework of accountability. This type of work may be difficult to assign standards or ROI criteria, but if measures of effectiveness are assigned to R&D it is much more likely that management will view them as productive activities and that they will be efficiently managed.

Let us consider the two facets of R&D separately. First, let us take up the question of what research should be conducted and how it fits into corporate strategic planning for the long term. Then we will take a similar look at development. Along the way the problem of budgeting and measuring the value of research and development will be addressed.

STRATEGIC OBJECTIVES OF RESEARCH

An I.E. might well view strategic planning as the construction of a tree of possibilities growing outward from the present. Each branch of the tree represents alternate future realities being chosen; each path from a fork on a branch represents a corporate strategy conditioned upon the existence of that branch's reality. In one sense, the anticipation of the branches and the forks is strategic planning.

A frequently important strategy in many cases will be to have a new product or service ready to fit an anticipated business environment. There is then a clear need to know what kinds of products and services the future might value. Toward this end a company requires constant market research.

MARKET RESEARCH

Almost every business is becoming more service-oriented, so it is preferable in this book to view end products of even a manufacturing business to consist of both goods and services rather than to constantly make the distinction. The long-term goal of market research is to predict the products that the company should develop for the future and their expected contributions to revenues.

These activities also influence short-term strategy and its information needs. A market research unit employed by a company should perform both short- and long-term analyses, the exception being the case in which outside researchers are hired for specific tasks. An in-house market research division is easier to coordinate into planning activities, but a hired service occasionally may be cheaper.

During planning, marketing projections must be filtered through two treatments of reality: (1) determination of whether the implied production or service levels forecasted are achievable: and (2) the assignment of costs to any proposed activities to discover their profitability. It is important to know whether expected future product markets can be entered profitably and competitively. The natural personnel to study capacities and assign costs are I.E.'s who have a solid understanding of the strategic importance of market research and development.

If marketing analysis functions are in-house, then I.E.'s should partake in the preparation of reports that forecast market trends and changes that might offer opportunities for profitable future enterprises. There is little point in reporting a rosy picture of an exciting new market without accompanying explanations of the costs and limitations of entering that market. Too often strategic plans are laid that insert a company into a cosmetic but detrimental endeavor.

The market research functions should be conducted by marketing personnel who understand the work, but I.E.'s will probably be the ones who attempt to put dollar figures on these research activities. Since market research is usually a staff function, it is ordinarily organizationally distant from line accountability and likely to have its costs subsumed under nonmanufacturing overhead. This approach to costing out research of such importance is an interorganizational mistake; hidden costs are always harder to control than they would be if they were made line items.

Research personnel of all kinds need to be convinced by I.E.'s that they will receive better support if they devise intelligible, consistent methods for costing their activities. One such accounting method is modeled on the research proposals that are commonly submitted by research organizations to funding agencies. Although it may seem strange for an internal unit to submit a proposal-like document to justify its own existence, if the unit is well understood by upper management, it is used intelligently by the company. If research is currently viewed with suspicion and receives insufficient support, formal proposals and budgets are tools available for educating management.

Consider the advantages. No mysterious costs will attach themselves mystically to market research activities; the research will include all costs broken out in a fashion consistent with the activity analysis of the firm. I.E.'s can help the researchers perform this cost accounting effort. I.E.'s can also help the researchers relate their activities to valid long-term strategic goals and objectives of the firm to ensure a matching of goals and the objectives as stated in the research proposal. Management will see that the researchers know what they are doing, why it is a good idea to do it, and they will see evidence that research can proceed in a businesslike fashion. These advantages should easily outweigh the restrictions that proposals impose on research. An additional benefit is that, perhaps for the first time, the C.E.O. can knowledgeably brag to his peers about what his research units do.

PRODUCT RESEARCH

Both pure and narrowly directed research should be continuously applied to products. Again this work does not have to be in-house. When it is possible to do so, management should consider asking for proposals from both in-house and outside sources to perform a specific research project. It may be found that outsiders can outperform insiders in which case several simple ways to cut costs will become immediately apparent. On the other hand, it may be found that outside labs tack on so much overhead to their real costs that inside work is far cheaper; this should be the case. In a well-managed company research work should only be let out when the company staff is swamped or when problems arise outside its expertise.

Lack of accountability coupled with lack of competition are the two reasons why any division of a company becomes a detriment. In fact a highly desirable procedure would be to have each unit justify itself via the proposal route, especially entrenched staff functions that commonly grow rapidly. Letting out requests for competitive bids can help to keep a research unit sharp.

The goals of product research are basic. They deal with the adequacy and limitations of products and with new ideas associated with the products of a company and its competitors. The work is an inextricable mix of pure and applied research and must be given considerable latitude for brainstorming. Designing safer products is a laudable goal of product research, as is improving products, simplifying products, and creating products with more flexibility of function.

Consider the simple flashlight; product research has produced flashlights to fit in shirt pockets, large flashlights, even those suitable for underwater use. Each kind has a market. Each came from solving a variety of research and development problems. Unfettered thinking is necessary to keep those bright ideas coming.

It is the need for unfettered pure product research that may incline management to assign a continuing sacrosanct budget to this category of investigation. This is a reasonable procedure even though it is an exception to the fully detailed accounting which a proposal approach would suggest.

MATERIALS RESEARCH

Materials research in many settings is difficult to isolate from product research. For example, many manufacturers are currently studying the use of plastic or ceramic parts where only metal parts have been used (e.g., plastic components of auto transmissions and ceramic turbine blades). These material substitutions can be viewed as mere product innovations and considered as part of product research.

However, materials research at the level of pure research does not resemble product research. Even the applications aspect of materials analysis and testing ordinarily seems so unrelated to products that many companies appoint a distinct scientific unit for materials work. The goals are to find cheaper, safer, or more reliable materials to accomplish existing or projected uses.

An I.E. trying to rationalize research objectives might be tempted to place dollar values on materials projects being pursued. For example, he might determine the potential benefit of a certain new material as a replacement for an existing one. If the cost advantage of the new material were not such as to ever repay the cost of the research, it might be determined not to pursue that line of investigation. This strikes at the heart of the crucial issue in research.

The issue is: How is research to be justified? The answer to this question in a company may determine how much will be spent on research, how much latitude for inquiry researchers are to be permitted, and how sure the benefits from projects must be. Two facts color the question: (1) research returns are highly uncertain; and (2) few new ideas will be produced without research.

The first fact means that research is sure to pursue some dead ends that return nothing for the effort expended. The second fact means that, if research is ignored, the company will forever be a follower of the leaders among its competitors. It is difficult to imagine an intelligent strategic plan that would accept such an obsequious role as a follower merely in exchange for the dubious benefit of avoiding the extra costs of research. An I.E. who finds himself in a company with a phobia against research had best reconsider whether to remain with it, because this company can be expected to oppose innovative ideas from him, too.

Research does create some managerial problems. Those managers who do not understand research tend to allow research departments to become autonomous. There is no surer invitation to laxity in the lab; an unsupervised lab is likely to work on useless, esoteric projects. Autonomy should not be awarded to any division at any level in a company, for there is then no way to coordinate it with the strategic plan.

The best organizational role for a research division is as a line function. It should be directly responsible to a line division and should have its budget approved on the basis of its research plan, which in turn derives from the corporate plan. Unfortunately, this alternative also presents the possibility that the line division will cut the research budget as a way to limit its total costs. This possibility should be restricted by injunctions in the strategic plan prohibiting such a move as unwise.

An I.E. trying to promote a rational strategic plan must make the organizational and philosophic needs of research clear to management, who must be convinced to explicitly support the need for research as part of the corporate culture. Some topics of research may be easier to promote due to their being easier to understand, such topics as inquiries into methods of production.

MANUFACTURING METHODS AND EQUIPMENT

Methods studies have long been assumed to be in the domain of the I.E., but as a research area it should be more encompassing, including the study of how equipment is used and even how equipment is chosen. Some of these choices are best judged by manufacturing engineers or other specialized production employees. Metallurgists, mining engineers, and electrical engineers are only a few of the other specialists that may be needed to judge the operational suitability of methods and equipment.

Few manufacturing firms make a continuing effort to study their methods and equipment; most such effort is expended when new equipment is needed. Then vendors will be invited to make presentations, and that may be the extent of the research. However, it is often difficult to interpret the claims of vendors. A continuing research effort would include budgeted trips by plant personnel to observe diverse equipment and procedures in action. Of special interest are how well they work, how much they cost to operate, what their maintenance records are, and the service record of the vendor.

An I.E. trying to justify budgeted expenses for methods and equipment research may meet resistance. Management is more inclined to allow this research area to be handled ad hoc. However, the I.E. should try to have this important topic included in the overall R&D strategy. As an argument for his advocacy, experience curves provide ready justification for methods research.

Experience curve theory holds that economic forces tend to reduce costs per unit of production as time goes on. Costs decline for one or more of several possible reasons. Either costs of materials are reduced, labor costs are reduced, capital-related costs decline, or operating costs go down. The labor learning effect is negligible after a short time, but major cost reductions should be attainable through material cost reductions resulting from better methods of production, cheaper materials, or the use of more cost-effective equipment.

These cost improvements are unlikely to be discovered without research, so an I.E. trying to justify the needed research budget has an excellent argument derived from experience curve concepts. A similar argument can be used to support research in quality control.

QUALITY CONTROL

Whether quality control more properly falls in the research or development category is an arguable issue and best ignored for the moment. It suffices that QC is an important, yet underfunded and poorly understood, aspect of management. W. Edwards Deming [9] cites management's lack of comprehension of this topic as a major reason American competitiveness has lagged behind that of the Japanese.

Deming should know, because for the past 30 years he has been revered by the Japanese as their prophet of productivity and quality control, the man for whom their prestigious Deming Prize is named. It is interesting to observe what Deming does when he takes on a company as a consulting client. His contention that management does not know the jobs that they direct is put into application.

He first analyzes the steps in production. Then he begins to focus on where quality control problems originate by studying the various jobs involved. Each job is characterized according to its normal operation and its statistical profile. Astoundingly, in most cases this is the first time this exercise has been performed in the company.

At last Deming has enough information to discuss where errors originate, how thay can be monitored and reduced, and, most important, what the jobs really consist of. Until management understands what the jobs are, Deming's suggestions will not do any good. How and how often to sample and monitor are the keys to quality control as taught by Deming. The results are fewer rejects, less rework, lower scrap quantities, better products, and happier customers.

An even more important result is that, when management fully embraces scientific quality control, workers must be educated to perform their tasks as system monitors, which in turn makes them feel that a better job is being done. Workers almost always respond positively to the opportunity of doing a better job and turning out a better product. Frederick Taylor [36], nearly a century ago, also held this to be true.

There is no need to bring Dr. Deming to every manufacturing facility in the country. Scientific QC is within everyone's reach and easily instigated by I.E.'s and QC experts. Capacity and activity analysis is the central work needed to define the jobs in a production process. Material standards and rework records should supply the clues that point to the high-priority QC jobs.

Often it is found that certain steps in a process tend to drift out of adjustment fairly quickly. These steps need to be equipped with simple indicators, such as control charts, that will indicate to the operator that QC is being endangered at that step. Educating the operator about statistical monitoring is an essential element of effective quality control. Sampling policy is another major part.

It can be argued that research into both quality control and methods improvement are so similar they should be organizationally combined. Certainly in smaller firms this is the usual result of limited staff sizes, but, if it is possible, it is better to have at least one QC professional to constantly monitor and study quality.

Recently many large American firms have become aware that their QC problems begin with their suppliers. They have ensuingly enlisted suppliers as partners in their QC effort. Toward this end, the larger companies have offered quality enhancement incentives, such as price schedules based on delivered quality of goods. This strategy has in turn led to the suppliers, perhaps for the first time, scrutinizing their own product quality. The chain of events is very healthy for American competitiveness.

Between working with suppliers, customers, and the production lines there is abundant QC work to be done. In addition, QC personnel should be integrated into all product development plans to elicit their suggestions early in the process. It is difficult to imagine how the lack of a QC research function could be justified in a manufacturing firm's strategic plan.

HUMAN FACTORS

Another corporate stepchild, often ignored, is human factors research. It is intimately connected with QC due to the fact that workers who become tired or who must work under poor conditions often perform poorly and contribute to low product quality. Human factors research studies how jobs should be done in order for workers to be able to perform them satisfactorily. Knowledge of factors that cause fatigue and injuries can be very important in reducing down time and low-quality work.

It is a sad fact that many companies devote little effort to studying ergonomic factors. One can visit assembly lines and see worker after worker forced to work in odd positions, employing heavy or ill-adapted equipment. The prevalence of poor working positions required to weld car bodies is one of the main reasons that robots are becoming widely used to perform this task.

A tired player is a coward, according to former Green Bay Packer coach, Vince Lombardi. Similarly, a tired worker cannot be expected to be at his best. Nor can a depressed worker. Correcting factors that cause fatique, discomfort, depression, and unsafe working conditions can be as important to a productive workplace as tight quality control procedures. In fact for labor-intensive operations the primary QC attention centers on the workforce.

As demonstrably important as well-designed work and comfortable, cheerful work stations are to productivity, many companies give these problems little more than lip service. The natural suspicion about a firm that does not show understanding of human factors is that it is run by people who are ignorant of the work environment. If this is the case then I.E.'s, who are educated to understand the basic need for ongoing human factors research, can try to educate the ignorant, although doing so may be difficult.

If a company is too small to have a full-time human factors expert on the staff, it would be advisable for it to plan on intermittently hiring a human factors consultant to study procedures and layouts. In addition, the firm should have its I.E. staff regularly interview personnel to discover their suggestions for how things should be done. Then the suggestions should be studied carefully in hopes of finding productive ideas.

Several of the cited research areas may have to be combined under the aegis of a single research office, but none of them should be slighted. Each can provide its own important contributions to cost control, innovation, quality control, and productivity. One as yet unmentioned research area that should receive constant attention is that which deals with the holistic design of operations and how they fit together, operations and organizational research.

OPERATIONS AND ORGANIZATION RESEARCH

Operations research (OR) is the discipline that studies operating systems. Its goal is to determine the characteristics of systems, describe them mathematically, and then determine whether they work in an efficient way. These principles can be applied to study the very structure of a firm.

An I.E. by training and experience is the logical person to take a generalized look at organizational structure. The questions to be asked do not vary significantly over time nor do the problems to be solved. Questions to be asked include:

Do corporate units have understandable budgets and goals?
Are there clear lines of responsibility and accountability?
Have units somehow become autonomous?
Does each unit fit the strategic plan of the firm?
Do new units have clear and distinct functions?

Negative answers to any of these questions suggest that organizational units may exist for no purpose. It is poor organization to permit a unit to exist that does not have line responsibility. A staff function responsible only to officers above it on the organization chart is probably divorced from line cost accountability. Such a function's effectiveness is difficult to measure.

Autonomous units are problematic because they cannot be controlled. A key to identifying irresponsible or autonomous units is that their budgets (if they exist) are not developed by systematic cost analysis. Such budgets are usually increased by an arbitrary amount each year. Units that have no separately identifiable budget are often supported by the ''slush fund'' of some other unit. A slush fund is a supply of money that can be spent arbitrarily. This type of budget chicanery in the federal government is called ''off-budget spending,'' a phrase remindful of ''nonmortal dying.''

Any budget that cannot be rationally explained is irrational. Any irrational spending is done in defiance of rational strategic planning and subverts it. This is one kind of problem operations and organizational research needs to detect early in hopes of eliminating it.

As strategic plans become operational in a company, rapid and precise information flows between organizational units become a pressing need. For example, the quicker production lets marketing and traffic know that orders have been completed, the better due dates can be met.

Designing and fine tuning systematic procedures that will facilitate rapid information flows is a major aspect of continuing operations research. This is the type of work which naturally calls for the strategically knowledgeable I.E., who is well versed in OR and who is capable of viewing corporate information flows holistically. Among the research aspects of R&D several important topics have been identified, but as yet little has been said of product development. Let us now concentrate on this topic.

DEVELOPMENT

Development is not always easily separable from research. In particular, long-range development projects invariably take on the appearance of research. However, there are instances when development stands out as a separate entity, as is the case in process development.

PROCESS DEVELOPMENT

The need for a specific process is often dictated by a company's desire to make a certain, well-defined product. The process may require a special-purpose mixing vat, a very hot kiln, or a variable-rate machining line. Even when a process is as well-defined, as, say, a patented one, it will undoubtedly require developmental work to fit it into the role it must fill. Even computer systems often require extensive effort to tailor them to the jobs they do.

Unfortunately, development is often slighted; at the completion of a research project specifications for a process are produced and research personnel pronounce the problem solved. However, nothing ever works in practice the way it has been imagined to do, so development work is always necessary. Although a firm will probably not require a separate process development division, technical personnel need to be available to do such work, and development costs must be budgeted for every new process.

MARKET DEVELOPMENT

Market development is not ordinarily confused with market research, but it may be thought of as a routine marketing function. It is true that it should be a job of the marketing staff, but it is not a routine activity. It's objectives are opening new markets and penetrating older markets more fully. Market development must be closely in tune with the corporate strategic plan.

A fad that companies adopted for a time was the installation of a Manager of Market Planning and Development, usually at the same rank as the Director of Marketing. The theory was that the market development job had become specialized, and there was also the feeling that interoffice competition might inspire Marketing to greater efforts. This is shabby thinking. A manager who senses intraorganizational competition is sure to think of little else.

Market development should not be divorced from marketing. Development is the result of the adaptation of strategic planning by use of research and is not an independent activity. Development plans should be based on directives from upper management that make clear the specific developmental goals to be pursued by marketing and explain their underlying rationale.

Market development, then, must be budgeted as a part of the strategic planning process within the marketing function. Product development must be planned and budgeted in the same manner.

PRODUCT DEVELOPMENT

Having new products ready to market when the time is right may be the most important strategy a firm can have. Often there is little research or deep thought required to identify a need, but actually producing the targeted product may require developmental skills that do not exist in the firm. For instance, for years managers have complained that there are no real computer *service* firms, i.e., companies that specialize in solving computer-related problems in the exact manner requested. It is thus made clear that a company marketing such a service would rapidly have abundant work. The constraint upon being able to offer the service appears to be the severe scarcity of personnel who can function with the mental flexibility and expertise required.

This one example of an as-yet-undeveloped service demonstrates the fact that a firm's development personnel must be entrusted to identify very carefully just how a product is to be made, priced, and marketed. Merely designing a prototype is not sufficient; many a military prototype goes "boom" prematurely. A genuine interlocutor between development and production personnel is necessary to monitor the progress and rationale of product development; an I.E. would be an excellent candidate for the job.

A frequent error that is made in product development is seemingly obvious yet committed constantly all the same; it is the error of short-circuiting valid product design and development. This is an error an alert I.E. can easily detect. Here is a common scenario for committing this error:

Marketing determines the characteristics of a product it has decided customers need and would buy, were it available.

The product idea is submitted to the design department, which draws up preliminary plans.

The plans go to the development division, which designs, constructs, and tests a prototype and progresses toward completion of the product's plans.

Marketing shows off the new product prototype to upper management in a presentation, stressing its sales potential.

Management buys the concept and orders the product to be made en masse.

For the first time the production personnel hear of the new product. Their response is probably one of:

1. We cannot make it.
2. We cannot make it for the cost assigned.
3. It will take us three years to learn how to do it.

The obvious question is: How does it happen that the people who actually have to turn out the product hear of it last? The answer lies with execrable management of the product development function. If a person knowledgeable of production processes were entrusted with oversight authority and told to coordinate the affected corporate units during product development, it is difficult to see how events could lead to such a ludicrous end.

The problem described is encountered with astonishing frequency, so there is a general lesson attached; the lesson is that development must be a highly coordinated activity. Its objectives should be to have products ready at predetermined costs per unit at the time markets are predicted to be ready for them. To be ready for sale and producible at sufficiently low cost are reasonable criteria for the product development function.

Many products are clearly needed and would sell in huge quantities if they were available at a reasonable cost. For instance, if homeowners could buy a complete home heating unit for $100 that operates year-round on solar energy, these units would generate enormous demand. The sad fact is that costs are not now low enough that a homeowner's ordinary heating needs would ever justify the cost of existing solar technologies. Cost is a major factor in product marketability.

Product development can be seen as an indispensible activity but one that must be monitored carefully to ensure that its efforts have a chance to enhance profits. In addition the products under development should be monitored to guarantee that nothing outside the corporate strategic focus is under development. An astute I.E., well versed in cost control, production processes, and strategic planning should be assigned to oversee this work. Almost the same array of characteristics are needed to supervise the development of standardization.

STANDARDIZATION

Standardization of processes, materials, and sequencing is a source of considerable cost savings. For this reason development of new methods of standardization should be a constant activity. Some aspects of this work seem to have been claimed by systems engineers and manufacturing enginers, so they may be the best professionals to direct part of it. It would seem, however, that I.E.'s should oversee the development of important measures of effectiveness, especially cost standards.

I.E.'s have long been noted for their standardization of work methods and their efforts in standardizing sequencing, scheduling, and other aspects of floor control, and these elements are still important to smooth work flow and cost control, but newer methods in areas such as machining make the use of specialized personnel in those areas desirable. For instance, the sequencing of parts through a numerical control (NC) machine should follow groupings determined by the principles of group technology, a manufacturing engineering topic.

Standardization is a discipline that is highly eclectic. As such it requires very specific assignments of tasks and budgets. The principle of standardization development is that no procedure is sacrosanct; all have room for improvement.

SUMMARY

All R&D is necessary, but not all such functions must be performed by separate company units. Marketing research may be done by the marketing department, and standardization work may reside with the various plant engineers. However, all of these functions are necessary and must be performed, which means they must be budgeted and must have specific, attainable objectives, attested to in writing by those responsible. The problem often encountered by R&D efforts is a lack of coordination of the diverse functions.

These functions involve diverse specialties, mostly technical disciplines. Yet to be meaningful each research and development activity must conform to the stated goals and procedures in the company strategic plan. Thus, R&D needs strong coordination effort if they are to be successful in reaching objectives and in adhering to budgets. Specialized technical personnel should be in charge of day-to-day R&D work and in initiating project ideas. But the individual who is best suited to designing, coordinating, and monitoring the R&D functions and who can best tell whether their costs and purposes are in line with budget and strategic plans is the I.E.

8
FINANCE

It cannot be stated too often—to a large degree budgeting is planning, at least for the periods budgeted. Unlike governments, a company must operate within its accessible resources. Thus the financial aspects of operations must be planned well into the future, as must contingency plans for shortages of funds.

Financial strategy has applications in both the long and short term, but the most important aspects of financial planning are inherently long-term. These aspects include budgeting procedures, finance policy, capital expenditure policy, and operating policies. As might be expected by the astute I.E., these financial planning issues rely strongly on the bottom-up analyses discussed earlier.

BUDGETING

Budgeting consists of planning how much is to be spent on identified corporate activities, where the money is to come from, and where cash will be acquired during cash-poor periods. Underlying the budget process are policy guidelines that detail which activities are to be line items, how much working cash is to be left on hand, how excess cash is to be stored or invested, what part of profit is to be retained, how large a dividend to pay, when to repay debts, and what needs might justify changing the budget.

Corporate budgeting might be likened to family budgeting except for the fact that a company will not often suffer as a result of miscalculations, whereas a family's members may go hungry. It is also true that corporate budgeting can be done very badly without the fact becoming public knowledge. Even if a company does make poor budgeting decisions, it may find friendly lending agencies or governments to provide support.

However, the degree of insulation from economic repercussions that companies can expect is not total, and many companies do go bankrupt weekly. It can be truly said that budgeting can be a corporate life-or-death matter. Yet the strategic principles of budgeting are straightforward.

First, only those activities that fit into the company's strategic plans should be funded. Second, only those funding levels that are rationally justifiable should be permitted. Third, only predictions of revenues based on factual analyses should be used to forecast how much can be spent. Fourth, an adequate operating cash allowance must be available at all times.

Consider the alternative. As soon as activities are funded that do not fit into strategic plans, proliferation occurs. Yet the likelihood of making an adequate ROI on a hodgepodge of unanticipated activities is nil.

If line items are funded at arbitrary levels there is never any basis for questioning how well the money is spent or for measuring the effectiveness of expenditures. However, many companies routinely increase the budgets of entire departments with no specific justification. The worst kind of irrational budgeting is that which permits activities to be paid for from slush funds, usually discretionary funds under the arbitrary control of a manager. A single slush fund may support rather large segments of a company, segments that have no stated purpose and no measurable output.

Even if outgo is rationally supported, a company can find itself in dire financial straits if revenues are not forecasted reasonably. There is a marked tendency in corporations to make pollyanna revenue predictions, and, once made, these predictions naturally lead to distributing unduly large amounts to line item expenditures in the budget. When revenues demonstrate a shortfall from predictions these inflated budgets must be cut in mid-term, producing difficult curtailments of some activities.

A company that does no budgeting at all, one that supports its actions in a "hand to mouth" fashion, will find itself short of cash often. Cash may not be available to accommodate accounts payable in a timely manner. The result may be a scramble to borrow funds. Budgeting should avoid this kind of problem by predicting when cash will be needed and planning ahead of time for how it will be acquired. A banker is much more inclined to look favorably upon a company that presents its cash needs in a well-reasoned manner with a long lead time than he is upon a firm scurrying about anxiously to cover current expenses.

These are basic considerations of budgeting, and they have motivated many recent managerial fads directed toward gaining rational control of budgets, cash flow, and financing methods. Zero-based budgeting, for example, is a method aimed at avoiding the automatic markup of the budgets of departments from year to year. The identification of strategic business units (SBU's) is an attempt to track the valid expenses of corporate units and assign measures of effectiveness to the way the money is spent. Experience curves are used to anticipate how costs can be reduced, which in turn permits reductions of budgets for the production departments. The activity analysis described earlier is also aimed at improving budget predictability.

There would not be such diversity of methods for budgeting improvement if the issue were insignificant. Even if one does not agree that planning cannot exist without concomitant budgeting, it would probably be agreed that rational budgeting is of value. Let us now focus on the aspects of rational budgeting and how they relate to strategic planning and to I.E.'s.

PLANNING AND FUNDING

Focus of activities must be a prime concern of the planning process, viewed here as a continuous evaluation and review of past and future policies. Not only money must be budgeted, but labor, facilities, and creativity must be channeled into a company's selected activity mix. As has already been discussed, the expenditure of significant capital resources on facilities or acquisitions necessarily directs a portion of the company's efforts. At the very least capital investments must recover their costs and show a return in real dollar terms.

Since only a narrow range of activities can fit into an intelligent focus of direction, only a small percentage of resources can be regularly permitted to partake in speculative ventures; the major part of resources must be dedicated to the mainstream activities adopted by the choice of focus.

If a full-blown activity mix has been projected for the long term, as must be the case when strategic plans have been laid, then the levels of planned activities will be a part of that mix's specifications. A given level for a given activity dictates a given amount of resource requirements for any upcoming budget period. If accurate understanding of activities has been acquired these resource requirements will be readily predictable.

For example, say that the company has decided to produce 50,000 units of its Jiko product next year. Assuming 50 weeks of production, that means that 1,000 units per week must be produced, if this level does not conflict with other scheduling considerations. If standards are available it should be easy to estimate the resources the 1,000 Jikos will require each week and thus easy to prepare monthly budgets. In this way the strategic plan for this product will create the budget requirements for Jiko's portion of the production effort. The calculations derive from the standards I.E.'s have obtained for the product.

Assuming that a company has planned its resources to be used in an intelligent manner in its strategic plan, there should not be large quantities of resources left unallocated once all components of the activity mix have been budgeted. Only contingency resources should be left over. Any directly budgeted funds not needed as working cash should be placed in temporizing investments.

The consequence of budgeting to fulfill the strategic plan is that there should be little or no resources left over to be expended upon arbitrarily chosen, spur-of-the-moment ventures. Spontaneity is the enemy of planning, and, while wildcatting can be profitable, no portfolio analyst will advise having a mix of resource investments that is predominantly speculative. This implies that intelligent planning leaves little funding for ad hoc investments.

The basis of proper resource budgeting is intelligently derived standards. A corporate activity without good standards cannot be budgeted rationally. A good example of a function that is commonly funded without rational underpinnings is that of the legal department. Legal work is naturally a staff function, a fact that automatically makes it difficult to assess.

The growth of corporate legal staffs is a recent phenomenon, apparently tied to corporate fear of federal enforcement agencies. Litigation with the FTC, OSHA, EPA, and other agencies can occupy large amounts of legal time. However, one might question the philosophy inclining a firm toward litigiousness. When the U.S. government insisted upon mileage, pollution, and safety standards for automobiles sold in the United States, Detroit sent their lawyers into the fray; the Japanese put their engineers to work and satisfied the majority of the standards by producing smaller cars in a limited variety of styles and colors. The Japanese appear to have spent their money on this issue much more intelligently.

Aside from questions of the wisdom of committing to legal confrontation strategies, a legal staff is difficult to budget because it is impossible to assign standards. By its nature it must continually create more work for itself, yet this work does not fit into a preordained plan, nor is it clear how to devise measures of effectiveness for money spent on litigation. A legal department is sure to have its funds budgeted in a wholly arbitrary fashion with no relation to the strategic plan.

An I.E. must be careful about the suggestion of the elimination of legal staff or of any permanent staff positions. It is sure to be a sensitive issue, but if I.E.'s do not take the lead in presenting ideas to save money, who will? The rule proposed is: If no rational basis for funding an activity can be given, its existence must be challenged.

The key to activity budgeting is standards. The total amounts budgeted in turn must depend upon income; no more can be spent than will be taken in as cash flow and additional preplanned financing. However, there is an unfortunate tendency to budget all income without regard to budgeted needs; in fact, staying within budget is generally difficult. It requires not only rational intentions but iron discipline as budgets approach their limits. The need to adhere to budget limits tends to make managers fear that their budgets may be too low, which in turn leads them to project overly optimistic incomes upon which to base their budgets. Whenever fictitiously high revenues are predicted, future cash shortages are made more probable; thus discipline must be employed in making revenue forecasts.

REVENUE ESTIMATION

Profits should motivate strategic plans. Activity analyses should lead to profitability analysis, largely under the direction of strategically versed I.E.'s. This analysis should form the basis for long-term and short-term strategic plans aimed at enhancing profitability. In performing profitability studies, reasonable predictions of consumer demands, product trends, price elasticities, product lives, and competitors' actions are needed. Much of this information must be obtained from the marketing department, from economists, from market research, and from well-kept company records.

Market prediction is exceedingly important; if a company aims at the wrong market or prices its products incorrectly, its demand projections could be much too high. If it badly misjudges the actions of competitors, the company's market share could also fall far below estimate. Thus major effort should be expended in preparing price, demand, and product projections.

Profitability analysis founded on well-researched data should be the corporate plan's foundation. Price times demand is revenues, so once price and demand are determined for each product, revenues have effectively been predicted. If all the research, statistical analysis, and study of competitors is accomplished scientifically, organized dispassionately, and incorporated into the profitability analysis by fully professional personnel, such as I.E.'s, then a rational basis for revenue estimates can be obtained.

Difficulties often begin at this point. In high-level budget conferences emotion and concern over loss of budget flexibility become sources of debate. This is a time when an I.E., dedicated to rational planning, needs to be present to defend the proposed budget. In the absence of an objective advocate a company may produce budgets as irrational and irresponsible as that of the federal government and with similarly disproportionate deficits.

Every department wants as large a budget as it can get. However, a department would not ask for more if the funds it obtains must be justified by its satisfying specified performance goals. Two factors are needed to prevent distortions of revenue estimates and the ensuing budgetary inflation: strong advocacy of rational budget projection and a firm hand governing the company. An indecisive C.E.O. can hamstring effective budgeting—he must be able to say NO.

Once a rational revenue estimate has been defended throughout the budget conferences, the company should be prepared to allocate funds based on standards. Revenue predictions necessarily derive from planned production levels. Based on the I.E. analyses discussed in earlier chapters, resource requirements to support those production levels throughout the company should be readily derivable.

EXPENDITURES ESTIMATION

Expenditures can be accurately estimated only for those activities that have been fully analyzed during an activity analysis. Problematic activities will not be line functions nor those budgeted in response to proposals but staff functions whose directors willfully resist standards development. This is the reason an I.E. performing activity analysis should use the opportunity to apply some method of cost allocation to staff functions far removed from line responsibility. Such allocation provides an arguing point at budget time for restricting the arbitrary growth of staff budgets.

Given a departmental budget that is fairly accurate in conforming to the company's needs, the management of that budget requires careful monitoring. There are two kinds of personnel who make the monitoring necessary: (1) the manager who is incompetent to manage a budget; and (2) the manager who believes he will have his shortfall automatically made up by the company and who purposely spends all his funds early.

In production departments, accounting personnel can easily track budget adherence because variances are recorded regularly, even daily, and a production manager will be informed promptly when he exceeds material or labor costs. The accounting department should similarly track variances in staff function budgets, but it cannot do so unless someone (perhaps I.E.'s) can present the accountants with a way to measure budget variances.

Again, it cannot be stressed too strongly that any department that cannot have its expenditures rationally estimated, related directly to the implementation of the strategic plan, and measured against expected levels of performance is a department that cannot be governed rationally. Its very existence is an invitation to arbitrary budget expansion and activity proliferation. I.E.'s must continually exert pressure to have cost standards applied throughout the organization.

For a very long time the old view of productivity, using units of production per labor hour as its measure, has predominated. Yet direct labor costs as a fraction of total operating costs have declined markedly since World War II. Staff costs meanwhile have expanded rapidly, as have capital costs. Now capital recovery and debt service costs dominate the costs of many companies. In consequence, a more valid measure of a company's productivity would be given by observing its total costs per unit of sales, a measure which finds direct labor costs to be a small component.

What this situation reveals is the existence of a management viewpoint that is stranded somewhere around the year 1900 when most American businesses were labor intensive. Present-day management needs to address its attention to monitoring so-called indirect labor and capital productivity.

As we have seen, the monitoring of indirect labor shows itself as a weakness in the budgeting process. If, as is common, staff functions constitute a high percentage of a company's costs, then inability to bring staff costs under control has dire consequences for the financial health of the company overall. I.E.'s are left with the responsibility to educate management to the modern needs of budget analysis. The second major cost component that has become steadily more important, that of capital cost and debt service, is also a matter for management attention.

CAPITAL EXPENDITURES

This topic was central to facility planning, disussed earlier, because no new facilities or facility changes can be pursued without the money to invest. Whether the money is available is a function of the company's financial health and its financial policy, and these are the topics that demand management's awareness of capital productivity.

Understanding of events always lags the events. Capital intensive operations have become the rule only recently. Thus budgets containing large line items for repayment of capital costs have only recently begun to creep into MBA textbooks, even though stable financial policies pertaining to investment have been needed for years. Most managers naturally learn about such topics as capital productivity on the job; others need help to grow into these concepts provided by financial personnel and I.E.'s.

Consider how capital investments are handled. A purchase of a building or piece of equipment is made as it is deemed necessary. The purchase enters into ensuing budgets in three ways. Periodic repayments of the cost of the facility appear as expenses, the interest appears as a deductible expenditure, and the deterioration of the facility appears as a depreciation deduction.

A common way to summarize the investment value of a facility investment is by citing its cash flow, depreciation plus incremental profit. This number reflects the funds liberated by the investment for further generation of profits. However, a better number might be retained cash flow, which is essentially cash flow less dividends paid out. Retained cash flow represents the two essential aspects of capital investments, their ability to produce an adequate return plus their ability to replace themselves.

It is an amazing fact that in recent years many companies have been paying out dividends to such an extent that their retained earnings were insufficient to replace deteriorating facilities. These companies were violating a cardinal rule of investment—they were consuming their capital.

A basic rule of thumb for capital productivity might be that the marginal retained earnings from an investment must be sufficient to amortize the facility's anticipated cost of replacement, including inflation and escalation, and after subtraction of that amount, the remaining retained earnings must satisfy the company's minimum acceptable internal rate of return. Let us say that a company relies on depreciation alone to recover the cost of replacement of plant or equipment. Recent decades have shown that the combined effects of escalation of construction costs and inflation of hardgoods costs exceed the capability of mere depreciation to support replacement. Yet many companies have persisted in paying dividends at a rate that has ignored their capital replacement needs.

A numerical example will clarify the point being made. For any capital expenditure, financial policy must dictate how the planned investment must function if it is to adhere to the tenets of the strategic plan. A simple rule would be that every investment must pay for itself in addition to paying for its own replacement while returning a reasonable profit over and above these requirements. Mathematically, we can write an equation for equivalent annual revenue requirements as follows.

$$\mathrm{R} = \mathrm{P(MARR)} + \mathrm{DS} + \mathrm{OC} + \mathrm{t(R} - \mathrm{EAI} - \mathrm{OC} - \mathrm{D)} + \mathrm{REP(A/F,i,n)} + \mathrm{DIV} - \mathrm{TC}$$

where:

TC = annualized investment tax credit
i = expected time value of money
R = annualized revenue requirements
P = principal invested
DS = annualized debt service
OC = annualized deductible operating expenses
D = annualized depreciation deduction
t = marginal tax rate
EAI = equivalent annual interest payment
REP = net future replacement cost
DIV = annualized dividend payment
$MARR$ = minimum rate of return on prinicipal

This equation attempts to summarize cash flow requirements into a fairly-easy-to-understand annualized revenue requirement. This equation is implicitly very conservative, since only that part of an asset that is purchased through outlay of principal is required to bring a return on investment. A more stringent criterion would be that both the principal portion and the borrowed portion of an asset should produce ROI. To clarify the use of the equation and to make the important points about ROI, let us consider a simple example.

A piece of equipment costing $200,000 is to be purchased by paying $100,000 down and financing the other $100,000 on a 10-year note at 20 percent. Let us say that the MARR for the invested principal is only 15 percent. Other assumed values for the equation are:

REP = \$200,000
DIV = \$5000
OC = \$5000

This replacement cost of only \$200,000 is a conservative value. It represents the anticipated replacement cost of the piece of equipment less the cash salvage value of the unit to be replaced. Recent replacement experience would suggest a much higher net replacement cost.

A marginal tax rate of 50 percent is chosen, assuming that marginal income is in the highest federal corporate tax bracket and that there are state and local taxes to pay in addition. For convenience the actual life of the piece of equipment is chosen to be the same as the loan payment period, 10 years.

The term, REP(A/F,.2,10), in the revenue equation represents the annual amount of earnings that must be retained in order to accumulate the cost of equipment replacement. Its value turns out to be \$7,704. Other values of note in the equation are:

P(MARR), the annual net profit on principal, is \$15,000
DS, annual debt service, is \$23,852
EAI, equivalent annual deductible interest, is \$15,710

Debt service is calculated by multiplying the original debt of \$100,000 times the capital recovery factor:

$$(A/P,.2,10) = 0.23852$$

yielding \$23,852. The interest associated with these equal payments consists of two parts, one constant and one changing as a geometric series over time. Specifically, the interest payment for year k is:

$$I(k) = \$23{,}852 - \$3{,}210(1.2)k$$

The geometric (second) term can be transformed into its equivalent equal annual series by multiplying the first number in the series by the annualizing factor for geometric series:

$$3210(1.2)(A/S,.15,.2,10) = 8142$$

Thus the equivalent annual interest deduction is:

$$EAI = \$(23852 - 8142) = \$15{,}710$$

An 8 percent tax credit the first year of owning the capital item was assumed, and this $16,000 was annualized by multiplying it by the appropriate annualizing factor:

$$\$16{,}000(A/P,.2,10) = \$3816$$

The depreciation term requires a brief explanation. It was assumed that the piece of equipment would be in the federal depreciation category for five-year depreciation, using the Accelerated Cost Recovery System (ACRS). Thus, the equivalent annual depreciation is five years' worth of depreciation spread over 10 years, assuming a time value of money of 20 percent. The resulting annual value is $11,276.

All of these numbers inserted into the revenue equation produce annualized revenue requirements of $73,494.

In effect, conservative assumptions have said that the piece of equipment must produce 73 percent of the principal invested in it intially per annum in order to satisfy the minimum requirements of providing a modest return, defraying costs, and replacing itself. The net after-tax profit required from this annualized calculation is $24,570, or 33.4 percent of revenues, of which retained earnings are $19,570 after disbursement of dividends.

Retained cash flow would be $30,846 and this is probably the most important number derived. It says that almost 31 percent of the principal value of the investment must be returned annually after dividends to permit minimal financial criteria to be satisfied. If larger dividends were paid out while annual revenues from the investment were only $81,126, there would not be enough cash flow left to amortize the equipment replacement, or else a smaller ROI would be realized. Either result would be a constraining financial policy.

What has been demonstrated is the stringency that sound financial policies impose on investment, ROI, and dividend decisions. The fact that management often violates prudent financial rules is well known. The question is why.

Some poor financial management is the result of ignorance, some the result of of poor judgment, and some the result of selfishness. Ignorance is a natural explanation simply because of the difficulty of analyzing investment information accurately. For example, as has been mentioned, only marginal revenues should apply toward investment justification, but marginal revenues are difficult to predict. Even more difficult to assess is the net replacement cost of a piece of equipment in the far distant future. Technological changes over time complicate this question, since there is no certainty that an identical replacement will be available.

Poor judgment is the natural name to apply to decisions made in willful ignorance or to decisions made in spite of sound advice to the contrary. It is a sad but true fact that many management decisions are made by people who have become caught up by the impetus of the moment. These people I.E.'s cannot help.

Selfishness may seem a strange motivation for poor decisions, but in recent years stock incentives have created this motivation. Say that a manager has stock in his company. He may even receive dividends just like any other stockholder. In any event, it is to his financial advantage to give out larger than justified dividends to reward himself directly through dividends or indirectly by inflating the market value of the stock. His assumption is that he will not be at his present post when it comes time to suffer the consequences of retaining an inadequate portion of cash flow.

The I.E. who is studying a planned equipment investment must restrict himself to the facts. However, he should also watch out for the abrogation of sound financial policy on the part of others in the firm and attempt to forestall them. If he spots a steady pattern of financial mismanagement, he should consider changing jobs before disaster hits.

The clear message is that financial policy must be part of the strategic plan of any manufacturing firm. Just as surely as a company plans not to bankrupt itself holding excessive inventories, it must also remember to sustain its investment in itself. It may be that many managers in this era of rampant acquisitions have lost sight of the fact that internal investment is the main goal of a viable company.

Investments in mutual funds or money market funds or speculative stocks may fit somewhere into the total financial plans of a company, but unless investments in itself dominate its portfolio it has no reason to exist. The law prevents a corporation from being no more than a holding company for money management, but there are ways around the law, such as acquisitions that actually represent nothing but paper entities to the acquiring firm.

All deviations from a manufacturing company's investing directly in itself are to be questioned. Few are justified. The internal rate of return of a manufacturing firm must exceed the return from temporizing investments, at least in the long run, or else the company is on its way to oblivion. A company spinning off its resources in such ways over protracted periods must be suspected of managerial cowardice.

It may be difficult to continue on a sound path of internal investment when market conditions are poor, and not all companies can survive. This is the Darwinian effect of the free market, but the company that purposely allows itself to be diverted from its own strategic plans is literally going nowhere. Avoiding this condition starts with sound financial policy.

Budgeting, rational revenue forecasting, and long-range replacement policy are building blocks of financial policy. Another is the day-to-day operating viability of the firm. Accounts must be paid and collected, and working cash must be planned and maintained if smooth daily operations are to be conducted.

OPERATING CASH

The level of operating cash a company requires varies sharply from firm to firm. If the sum is quite large, most of it should probably be kept in a liquid investment account. On the other hand if the amount is small and turns over frequently, an ordinary bank account will suffice.

Most companies budget cash by quarters, so it should not be difficult to plan ahead and ascertain quarterly cash needs. If the strategic plan is on course, and if expenditures have been put on a rational basis, cash needs for the next several quarters should be easy to project. A reasonable delay can be built into retiring accounts payable, and past history should reveal the lags to expect in accounts receivable being paid. Sales revenues should be similarly estimable. Given these estimates, most expenditures are sufficiently regular that, except for emergency costs, cash needs should be able to be budgeted for several quarters.

A projected schedule of income and outgo make it possible to predict when the company would be strapped for cash. A simple rule for operating cash might then be proposed such as: Avoid being out of cash more than five percent of all quarters. Simple statistical analysis and cash flow simulation could then produce an estimate of the required cash level.

Since holding substantial amounts of cash on hand is counter to the usual demands of strategic plans, a company would ordinarily prohibit holding an excess above the estimated amount. This prohibition naturally leads to being short of cash occasionally.

The beauty of rational budgeting is that shortages are not problematic when they are known beforehand and planned for. A lending institution presented with a sound budget will not object to giving a company a standing line of credit if the company has demonstrated the ability to understand its own finances. Many expected cash deficits do not occur due to the fact that financial officers are able to manipulate funds in the short term. This ability to manipulate is abetted by keeping a working cash level that is very close to the correct one.

An I.E. working with management on financial policy may have some small difficulty explaining the idea of working cash levels based on statistical criteria, but success in this effort will help promote smooth operations. Since it is understood beforehand that some extra funds will be needed from time to time for a variety of reasons, this issue also must be rationalized as a part of financial policy.

OUTSIDE FINANCING

Short-term shortage of cash is a minor problem that can easily be solved via short-term debt. This is one of the few valid uses of such debt. However, for long-term purposes funds must be raised from time to time, and for these purposes there are two main funding methods: stocks and bonds.

There are advantages and disadvantages to both long-term funding methods. Selling stock obligates the company to consider the welfare of its new stockholders and to submit to public accounting practices and SEC regulations. Long-term debt in the form of corporate bonds requires regular payments of interest and an eventual return of capital.

On the positive side stocks do not impose specific dividend levels and do not necessarily form any marked constraint on operating philosophy. Stocks provide capital in large amounts when needed for healthy companies. Corporate bonds make no demands on corporate policy, and a healthy firm should have no difficulty meeting interest and principal payments. In addition, the cost of capital raised by selling corporate bonds is theoretically less than the cost of stock financing [5].

Whatever modes of financing are adopted, they should be determined long before the need is at hand. Many a company finds itself having to sell off some of its best assets when it discovers it cannot obtain long-term funding when needed. A vicious cycle begins when a company goes into debt to finance its needs and then cannot pay the debt service and must go further into debt simply to operate from day to day. This is a truly unplanned existence.

There is another advantage to issuing stock. Stock can be used as part of an incentive program for workers and managers. Issued as a part of a profit-sharing plan, stock can encourage workers to take a stronger interest in the success of their employer. It is, however, more common for managers to be given treasury stock, and this practice holds the possibility of abuses, some of which have been mentioned.

Unless a company is going to embark on massive new enterprises, it may not be a good idea to think of financing activities exceeding its own retained cash flow. Companies are known to have opted for outside financing for spurious reasons only to regret it later. Whatever financing is used, it must fit into a well-conceived strategic plan.

SUMMARY

I.E.'s can play a central role in financial planning. Since this part of strategic plans must be based upon accurate information, intelligently conceived forecasts, and operational realities, it seems natural that I.E.'s should be involved in several aspects of rationalizing finances.

The budgeting process requires simulation of future costs and revenues. These values must be based upon broad perceptions of how the company really works. An I.E. can be of great help to planners by preparing well-designed simulations producing workable budgets that avoid the error of not planning properly for strategic needs.

A common corporate error is made by not questioning whether poorly understood budget items are cost-effective. I.E.'s can direct their attention at areas that have not received serious budget scrutiny, such as staff expenditures. At the same time they can develop standards against which performance can be measured.

In addition to providing an improved understanding of revenues, costs, and budgets, I.E.'s can call attention to real cash flow needs motivated by facility requirements. They can alert management to schedules of equipment and building repair and replacement that can then be properly considered in selecting investment alternatives.

From time to time specific projects, such as feasibility studies will probably be prepared by I.E.'s to assist in choosing between alternatives. In summary, I.E.'s should be entrusted with performing feasibility studies, coordinating cost and revenue projection models, monitoring the supporting data systems, instigating cost control, and studying the financial aspects of production.

To do a fully comprehensive job of these strategically important tasks, an I.E. involved must understand the strategic policies upon which financial policy is based. This requires that financial policy be explicitly established and that its I.E. be fully briefed on that policy.

9 SUPPLY SOURCES

Sources of supply have been a central topic in the literature of war strategy since the days of the Roman legions. In the competitive realm of manufacturing businesses, timely, accurate quantities of materials and parts are as vital to corporate survival as stores of grain were to Julius Caesar. Available strategies begin with securing the best supply sources. The essential distinction between rational strategies lies in the choice of the criterion used to define "best."

One strategy is to ensure that materials are available from a multiplicity of sources. This strategy has long been a trademark of U.S. manufacturing concerns. Another strategy is to develop a small number (perhaps only one) reliable supplier with whom to work closely for especially accurate and timely supplies. This approach closely imitates that adopted by large Japanese firms. A third strategy is vertical integration in which every stage of production is in-house, such as at Henry Ford's Rouge plant.

Multiple supplier strategy satisfies a criterion of low risk of supply interruption. Quality assurance and inventory minimization are the goals of the the Japanese approach. Independence from suppliers and control of costs and quality seem to direct vertical integration.

Each strategy includes a set of tactics for reaching its goal. If the strategy has been rationally developed the tactics will make logical sense, but if the strategy has been haphazardly adopted the tactics will probably seem self-contradictory. Unfortunately, once procedures have existed in a company for even a few months they become hard-to-break habits. The job of the strategically aware I.E. is to try to bring rationality to the exercise of supply decisions. Toward this end he must first determine what supply objectives fit into the company's overall strategy.

SUPPLY OBJECTIVES

The grand strategic objectives shaping supply decisions descend from the end user of the product. If a manufacturing firm intends to sell a commodity item to a broad spectrum of customers, then the ability to produce large quantities of product predictably will guide the supply choice. If a select, tightly targeted clientele is to be served then product reliability will be of prime importance and should motivate supply sourcing. Quality may be most important even for a commodity item in which case both quality and quantity considerations must be served.

Operational objectives must also be factored into supply considerations. Schedules, inventory policy, and cost control can all play a role in determining sourcing strategy and tactics. The control of both material quality and product quality is an important aspect of production and may dominate supply strategy.

If an I.E. is to make sense of supply policies, he needs to make quantitative evaluations of alternatives wherever possible. Some approaches to doing so follow, identified by decision categories.

SUPPLY RELIABILITY

Accurate scheduling of routine work is essential to cost-effective operations. In most modern manufacturing facilities, product schedules are exploded backward from due dates to subassembly, part, and raw material stages. The various levels of product assembly are sequenced so that order quantities and order releases can be achieved as scheduled.

Scheduling accuracy requires that lags between orders and deliveries be predictable. For instance, the scheduling algorithm may project that the due date for an order can be met if materials can be obtained from a supplier within a week after they are requested. If the lead time assumed for deliveries from that supplier is one week, the manufacturer will plan production based on this fact and assure his customer of delivery by his due date.

Two problems commonly arise to disrupt timely deliveries:

1. the supplier is unable to send the supplies;
2. the supplier displays erratic lags between order and delivery.

Tactics that can be tried to improve lead time reliability include: securing several suppliers; holding buffer stocks of critical parts; and writing price penalties for late deliveries into purchase contracts.

Ordering supplies to fill a specific order is only one of three conditions that triggers orders of supplies. Routine work requires steady and fully predictable deliveries and must be viewed as possessing distinct supply requrements. Yet another unique case is that of the rush, or expedited, order.

We can consider these as three different situations for which different tactics are required. Situations demanding special orders may occur often at a job shop facility and occasionally at a smooth-flow facility. Lot-for-lot ordering of supplies to fill orders is the usual mode of operations at a job shop. Routine production, sufficiently repetitive to be fully predictable, can occur at job shops, smooth-flow operations, and assembly lines. At the plants that sell smaller quantities per customer the routine product runs are probably directed to filling inventories, while they are almost the only kind of run of an assembly line. However, even an assembly line may occasionally need to order an unexpected quantity of supplies. In any of these instances supplies need to be reliable, so tactics to improve their reliability need to be employed.

To an I.E. monitoring the workings of orders, schedules, and deliveries, the important property of lead times is their accuracy. It is a surprising fact that many MRP systems have been set into operation containing wholly unreliable lead time data. Since the accuracy of recorded lead times is critical to the meeting of promised dates for finished products, the I.E. must carefully monitor observed lead time variances to determine whether they are significant. If recorded lead times are not right, they are sure to be longer than expected (a corollary of Murphy's Law), and corrections must be made.

In the case of a single supplier the response might be to simply revise the lead time upward. Or the response may be to seek alternate sources. If multiple suppliers are used as sources, it may be that only one or two are significantly late. These suppliers can be abandoned as a corrective action. Contracts that include premiums for adhering to lead times and penalties for being late can be employed as carrot-and-stick approaches to obtaining supplies on time.

In the case of routine, repetitive production runs, the supplier knows exactly how many items are required and exactly when they are required. Assuring reliability of supplies in this case is very important, and it might be advisable to leave the responsibility with a single supplier in order to make this segment of his business large and important to him. Distributing the total demand for supplies over several suppliers is a possibility, but probability theory indicates that deliveries are sure to be inaccurate more often in that case.

A natural response to the expectation that supply deliveries will sometimes be inadequate is to maintain inventories of buffer stocks. This alternative will be specifically addressed below. However, buffers are more appropriate to job shop facilities than to those performing routine or intermittent production. However, buffers may be required at any plant that expects rush or expedited orders.

Rush orders or expedited orders are usually of two types: orders for high-priority customers, or orders that have somehow become perilously late. Neither mode of expediting is a cost-effective activity, and a prevalence of these kinds of orders is a sign of serious mismanagement. The frequent effect of hurry-up orders on the supply function is that expedited deliveries must be obtained at a high price. If we grant that some such work is unavoidable, there are several approaches ordinarily available for acquiring extra supplies post haste that will be less costly.

One approach is through a cooperative agreement with another manufacturer in the same line of work as your own company. Consider refractory manufacturers as an example. Their plants are often near one another because they use similar mined raw materials. Yet each plant imports special supplies from distant places such as South America, which require long lead times. It is not uncommon for such plants, even though they are owned by different companies, to have mutually satisfactory ad hoc agreements to provide interim stocks to one another in the case of shortages of supplies, such as South American alumina.

Another way to satisfy expediting needs is to use local suppliers, who may be happy to cut prices slightly as encouragement to an infrequent customer. In the refractory business there are many small refractory plants near steel mills, so if a steel mill finds itself with a sudden need for plastic to line a soaking pit, it may well contact one of these small, nearby refractory suppliers.

A final method of providing rush supplies is to produce the needed materials in-house. This tactic requires a considerable commitment by a manufacturer to an expansion of activities and runs the danger of distorting its focus of activities. Even small-scale vertical integration impacts several categories of strategic policy, such as facilities and financial decisions.

As a major tactic in a firm's supply plans, in-house operations should make supply schedules easier to predict; costs should be easier to control; and quality should be more readily assured than in the case of outside suppliers. Obvious problems attendant to vertical integration have already been discussed in prior chapters. The tradeoffs between outside and inside suppliers must be analyzed factually before an intelligent choice can be made, and it will ordinarily be the province of an I.E. to prepare the economic and operational evaluations.

Major considerations in choosing supply procedures are quality assurance, scheduling flexibility, adherence to inventory policy, and cost control. Brief looks at each of these aspects of supply needs follow.

SUPPLY QUALITY

Quality of supplied parts and materials is a major determinant of final product quality and of the amount of scrap and rework generated during manufacturing. Recently it has become more common for large manufacturing firms to establish quality criteria for supplies. A good example is the automobile industry.

In response to the public perception that their U.S. cars are of lower quality than Japanese cars, U.S. automakers have adopted a strategy of improving quality. Toward this end they have insisted on higher quality supplies from steelmakers. In turn, it has been discovered that some of the quality standards had never been well understood or had never been studied by some of the steel companies.

The automakers have chosen several responses to the steelmakers' lack of quality control expertise. They have applied price penalties to lots that are below standard, e.g., sheet metal that deviates too much from uniform thickness or coating. They have also adopted the innovative tactic of working *with* the steel companies by providing them with technical assistance in the form of metallurgical technicians and quality control specialists. Ford has chosen to maintain a partial vertical integration by producing some of its own steel.

In the search for quality, low-quality suppliers will eventually lose business. A cold-blooded view of economic theory holds that poor suppliers *should* fail. In a shrinking industry such as steel, some suppliers must go out of business, and quality is an efficient selection criterion in the Darwinian process. The lesson seems to be that in highly competitive industries quality must become an important strategic consideration.

To the I.E. hoping to help assure quality of supplies and of products the difficult question is how to detect and monitor variances. A seemingly trivial point, but one of the most common failings of manufacturing firms, is their inability to keep track of scrap. This is a routine shortcoming of all kinds of production facilities, and it complicates the I.E.'s job. For instance, a printing plant finds that it is running out of paper much faster than its production schedule would predict. A foundry is unable to discover where many tons of the metal it pours has gone; the scrap rate is too high. A steel mill finds that final stocks from production runs are far short of planned levels.

Numerous other examples could be cited; the problem is pervasive. The point is that when scrap cannot be traced it cannot be attributed to specific causes. For instance, is the high scrap rate at the printing plant due to operator incompetence, excess moisture in the paper supplies, or the kinds of jobs being run? If an adequate information system to answer these questions is not in place, the sources of quality variances cannot be detected.

Consider the plight of the I.E. trying to improve quality in the absence of adequate data. In trying to determine whether a supplier is sending substandard materials or parts he may not be able to make a decision. A small metal fabricator that makes a variety of metal devices from sheet steel found that testing of the sheet metal upon arrival from the steel mill did not predict the production scrap rate. It was eventually learned that the grain structure of the metal had gradually been changing, making the metal harder to draw without cracking or breaking, and this problem created unexpected scrap in the cold-drawing operation. Only after the location of the excess scrap was found did detailed metallurgical analysis of failed materials reveal the source of the difficulty.

Not only do some quality control problems conceal themselves like the one at the metal fabricator, other sources of scrap and reject variances mask quality deviations. Yet these deviations are exactly the flags indicating that supplies are not up to standard, and they need to be detected in sufficient detail that erring suppliers and bad lots of materials can be identified. Yet the tracking of materials requires an uncommon attention to detail.

Someone must design the information system that can permit quality to be tracked properly, and an alert I.E. seems to be the ideal person for the job. Virtually every manufacturing facility that has accounting standards for cost tracking can over time discover material variances, but not many such facilities can readily identify the source of the variance as being a specific supplier and supply order, yet, this is the detailed identification needed for quality control.

In short, each job in a production process must be sufficiently well understood that each source of major contribution to reject rate, rework rate, and scrap can be observed. The development of activity standards mentioned earlier is a first step to this understanding, but, for the purposes of Quality Control monitoring, a statistical profile of sources of variances must be developed. Adequate information must be routinely accessible that will permit intermittent observation of variance components. Then, if a supplier is found to be frequently guilty of poor quality, remedial actions can be taken.

Complete understanding of production steps that incur material losses can only be obtained by thorough observation and scientific statistical analysis. The best approach to studying critical production steps is to develop an information gathering and information analysis system that is designed for manufacturing control rather than for the accountant.

The difficulties attendant to discriminating between sources of materials problems can be seen from a small example. Say you expect an order for 1000 gizmos to require 10,000 pounds of steel strip, based on a material standard of 10.0 pounds per gizmo, but after the 1000-unit production run you discover that 11,200 pounds were actually signed out from material stocks for this order. The question is what caused the 1200 pound variance.

Suspected sources of overrun are employee error, machine maladjustment, and poor metal quality resulting from inadequate supplier quality control. One cannot reconstruct the causes of excessive scrap after the fact unless information systems exist that can provide the needed discrimination. A simple detection procedure in this case might consist of saving the scrap from each run in a scrap bin for that run until that scrap has been inspected.

If the causes of scrap are readily identifiable from the nature of the scrapped pieces, then that knowledge plus the use of the single scrap bin would allow for a complete analysis of that run. Perhaps operator error shows up as improper cutting or off-center pressing. Machine variances may leave streaks or inhomogenous thickness in the metal. Broken metal may be the clue to structural weakness resulting from metal of poor quality.

The inference we can draw from this example is that to track supply quality variances the person in charge must know how to identify poor quality and must have the quality-checking criteria developed and the information systems installed that can relate material variances to poor supplier quality control. Often the information system consists of an acceptance sampling procedure that analyzes each batch of materials received. Acceptance will be based on statistical quality criteria, and the sampling procedure can be no more than minimally destructive if the sampling is to be a cost-effective approach.

The installation of a proper information system for QC analysis should naturally involve QC experts, production managers, and I.E. personnel. Only a fully integrated effort, guided by strategic criteria, can be expected to succeed.

One of the goals of material quality assurance is to provide for predictable production; low-quality materials imply higher material consumption per good unit of finished product, which also implies slower production. However, variations in all aspects of production are to be expected even under the best conditions, thus some flexibility must be provided.

FLEXIBILITY

Say a wood products factory runs out of a needed stock of wood. This will happen occasionally even in properly managed factories, so there should be a contingency plan for acquiring emergency supplies. The plan should access supply more quickly than usual. Every firm should have the flexibility provided by a plan for acquiring rush supplies, but this plan should not be implemented frequently.

If we visualize supplies as being of three kinds, we can class emergency supplies as being one kind, but they should constitute only a small part of total demands for supplies. The other two kinds of supply quantities are those required on a much more predictable basis: (1) in response to long-term production plans; and (2) in response to final orders. Long-term, consistent supply orders are generated by production plans to fill consistent, routine, long-term product demands and to stock inventories. Supply orders generated by firm customer orders should ordinarily have adequate lead times that no emergency orders will be required. Orders generated by long-term demand and firm orders should constitute the bulk of supply demands.

If a firm finds that it is constantly acquiring short orders of supplies in the spot market, this is not merely a sign of its need for occasional flexibility of supplies. Instead, its scurrying about to maintain operations shows it to have poor supply policies. These policies should descend from the strategic plan, which in turn, should be founded on solid bottom-up analysis. A firm with well-understood lead times and scheduling procedures should not find supply flexibility a frequent need.

Just as the handling of order quantities strongly influence supply orders, so does the handling of inventories of final products and supplies. Confused inventory policy is sure to confuse supply policy.

INVENTORIES

The classical inventory policy that has received lip service for years, the economic order quantity (EOQ) approach, seeks to balance the costs involved in ordering supplies against the cost of holding surplus materials in inventory. In the case of final products, holding costs are compared with the setup costs required to initiate a production run.

This approach assumes there is an unquestionable need for buffer stocks and that highly undesirable consequences accompany running out of stock. However, in recent years the maintenance of extensive safety stocks has become a liability as holding costs have risen. Notable alternatives include the use of just-in-time scheduling techniques such as the Japanese kanban and the American closed-loop production control systems.

Recent inflation and high interest rates have impugned the basic assumptions of the EOQ approach to inventories, since holding costs have come to far outweigh setup and order costs in most industries. Shorter production runs and smaller lot sizes are therefore becoming more common. Government economists believe data show total U.S. inventories to have become permanently smaller as a result of this trend.

MRP II (a short title for Manufacturing Resources Planning) is a systematic approach to scheduling all aspects of resource utilization at a production facility, and when it has been installed successfully, it reduces the magnitudes of the uncertainties that previously inspired the holding of safety stocks. The net effect of an MRP II system on inventory policy is to reduce safety stocks and schedule orders and deliveries more tightly. The impact on supply policy is significant.

For instance, consider a little publicized aspect of MRP scheduling systems. Most manufacturing firms at some stage in their processes hold stocks of materials or parts that have diverse uses. Since these items are ordered so frequently, they tend to have unpredictable levels, which makes them act as though they have independent demand. These stocks, which we can refer to generically as "intermediate stocks," will accumulate, creating inventories even in a tightly scheduled system whose aim is to eliminate inventories.

In particular, the kanban system requires that materials be delivered to the production facility "'just in time" for its use in a scheduled production run. The avowed purpose of this system is to eliminate inventories, and in fact, kanban plants maintain buffer stocks nowhere within the plant. The reason is that the intermediate stocks have been moved all the way out of the plant into the warehouse of the supplier, who must maintain and pay for them.

The peculiar aspect of this transposition of safety stocks is that holding costs must still be paid, but they now no longer appear as direct cost in the manufacturing plant records. Another aspect of this peculiarity that deserves attention is that these safety stocks may not be managed economically. As a result of their certain existence and their need to be paid for, a just-in-time manufacturing firm would be wise to work with its suppliers to help them reduce their stocks just as it works to minimize its own.

If a firm decides to minimize stocks of materials, it must enforce stringent criteria on the timeliness of deliveries from suppliers. Otherwise, material stocks will accumulate as a result of the variability of delivery times. The fact that deliveries are not timely is another piece of information that the I.E. in charge of information systems must be sure to obtain, and this will require coordination between the I.E. and the personnel responsible for delivery acceptance, warehousing, traffic, and order entry/release.

If minimal in-plant material stocks are to be held, then coordination with suppliers should be organized to make the supply function a symbiotic activity that is efficient, timely, and cost-effective. Toward this end, adequate communications between user and supplier firms must be established.

If final product stocks are to be kept in abundance, schedules for their production do not need to be tight, which in turn, means that no pressure need be applied to materials suppliers. However, if final stocks are to be kept small, as in the case of a plant that produces only to fill firm orders, then time pressure will be felt at all stages of scheduling all the way to the supplier.

In summary, inventory policies implicitly dictate the way suppliers are handled. More importantly, inventory policies dictate to a large degree how a plant should be run if its operating procedures are to contribute to its strategic objectives. However, the application of any supply strategies should also satisfy the dominant criterion of cost-effectiveness.

TOTAL COSTS

Who better to monitor and analyze supply costs than the I.E.? The permissible supply alternatives should be analyzed for their cost implications. For instance, if vertical integration is contemplated, it should be justified on a basis of providing design flexibility and ensuring material availability while lowering total production costs.

Costs of vertical integration include the high cost of capital recovery plus the operating costs that go with additional production. If internal supply of materials is found to be the low-cost alternative, then the other supply modes must indeed be expensive. It is not hard to visualize internal supply being justified for noncost reasons, however. The ability to adjust designs and material specifications rapidly is attainable almost exclusively in the context of internal supply. Another motivation is that owning one's own raw material supplies is sometimes the only reasonable way to reduce the danger of supply interdiction.

Whatever the choices being considered and regardless of the ultimate selection criteria, the total costs of alternatives need to be estimated during the selection process. The costs assigned to a supplier should include the estimated expense of rejects and rework. If the choice of supply mix creates more or less need for holding buffer stocks, this fact should be included in the form of a cost estimate, too. Scheduling problems brought on by late deliveries and short lots incur opportunity costs that must also be estimated and included in the cost per supplier per unit of materials.

Note that opportunity costs are not real costs. None of these costs in fact must correspond to the kinds that accountants commonly record. Their purpose is to provide a basis for estimating the cost effects of suppliers, even when costs are influenced by interaction effects.

Prior to selecting a mix of supply sources or a single supply source the costs cited may be little more than guesses, but, as experience mounts and data accumulates in information files, improved estimates of costs can be associated with suppliers, allowing periodic reevaluation of sourcing choices. This recurrent exercise is typical or the review and evaluation procedures that need to be performed to rationalize and monitor strategic policy implementation in all facets.

SUMMARY

Sourcing of supplies is important to several strategic goals: quality control, reliability of material source availability, production flexibility, inventory management, and cost control. Whatever sources are chosen should be justified by how well they satisfy strategic objectives. Toward this end the supplies from each source should be monitored by I.E. personnel to determine their supply lead times, their delivery accuracy, their quality, and the total costs per unit associated with each source. In addition, supply sources must be developed to satisfy each of the long-term, intermittent, and rush needs.

The need for continuing review and evaluation of sources in turn requires that information systems be established to provide the data for analysis. I.E.'s will probably be the personnel to design, install, and monitor the information systems that facilitate tracking of scrap, quality control problems, and erratic suppliers.

10
DISTRIBUTION

Chapter 9 dealt with bringing in materials from diverse sources. This chapter looks at the mirror image of that problem—distributing products to diverse end users. Again, there are numerous strategic and operational aspects of the decisions that must be made.

STRATEGIC ASPECTS

Service levels and costs are the dominant strategic issues determining distribution policy. Commodity items must be delivered in large numbers to sites of final demand as required or risk a loss of market share. However, a single-minded emphasis on avoiding lost sales can lead to the adoption of a costly policy of overstocking, which fact demonstrates that good service tends to incur costs by affecting inventories.

In addition to supplying customers with their routine needs, strategy may call for stimulating demand through promotions, sales, and special deals. Successful sales promotions require the ability to satisfy wide swings in demand, which in turn suggests the need for flexibility in the levels of units available for sale. Such flexibility may be obtained by juggling supplies of products from storage site to storage site, or it may be obtained by revising the way production runs are scheduled, or, as was suggested above, it may be obtained by overstocking.

If cost control is the strategy to be followed, then one sure goal is to keep a minimum of finished goods inventory. Further, a minimum of interference in orderly production runs would be permitted. Cost control as a dominant motivation tends to minimize other issues, such as avoidance of stockouts, so cost control may seem to conflict with service and promotion objectives.

It may be that a company would choose to balance the considerations of service, flexibility, and cost in adopting a distribution strategy. Also to be considered in this decision are decisions pertaining to other strategic plans within the company and the impact the adopted plans have on operations.

OTHER STRATEGIC CONSIDERATIONS

Distribution plans cannot be made in a vacuum; they will interact with all other policies that have been discussed throughout this book. In particular the modes of distribution affect facilities decisions, the company's strategic focus, its need for acquisitions or expansion, and its marketing policies.

Basic distribution decisions include whether to sell directly to consumers, whether to sell directly to retail outlets, whether to operate retail outlets, whether to sell to wholesalers, what transportation methods to use, where to locate warehouses, and how and when to provide stocks at termini throughout the distribution network.

The decisions adopted in response to these options build a set of distribution tactics that must be consistent with the company's broad strategies. For example, acquiring new warehouses in faraway places must be tested against investment and marketing plans to determine the wisdom of such moves. After strategic conflicts are resolved distribution plans must be tested against operational realities to determine whether they can actually be implemented.

OPERATIONAL ASPECTS

As will be discussed in more detail later, any plan that requires manufacturing functions as part of its resource needs must pass the tests of feasibility, economic reasonableness, and possibility. Feasibility is established if the modes of operation required and the levels of output needed can be achieved. Economic reasonableness is demonstrated if estimations of costs incurred by these operating modes do not exceed limits imposed by profitability criteria. Possibility is satisfied if the entire system within which the operational activities are intended to be coordinated can be shown to be workable.

For example, automakers can design makeable new cars; this satisfies feasibility. They can price them so that they would be profitable to make, thus showing them to be economically sound. However, if an adopted design would require large changes annually, this would not be possible. It is very important that the question of possibility be fully explored prior to the adoption of new, untested procedures.

It is a good idea for an I.E. studying new distribution procedures to use mathematical models to simulate plant operations under contemplated new policies to develop a picture of plant flows, input and output rates, costs, and schedules that would result. This work can be done with spreadsheet software, simulation models, or previously developed activity analysis tools. This type of study must be entrusted to someone such as an I.E. who understands the overall operating environment, the strategies underlying distribution policy, and the operational considerations peculiar to distribution.

DISTRIBUTION

It has become common in recent years to view distribution as the front end of a system that takes an input flow of materials and other resources through stages that transform the inputs into finished goods. Models for this visualization of the production function are sometimes called manufacturing resources planning (MRP II) systems, and the module at the front end of the computational model is the distribution requirements planning (DRP) module. In effect, this part of an MRP model is a computational algorithm that takes final demands and maps backwards to the resource requirements necessary for filling them on time.

The schedule of final demands may consist solely of forecasted demands, solely of firm orders, or some mixture of the two. Demand may be given as a single number, or it may be broken out into regional or store by store demands. After a schedule of demands at final locations is obtained, the DRP module explodes product requirements backward, developing a production master schedule and attendant inventory levels.

The DRP may then iterate with an inventory policy module to determine whether the master production schedule must be revised and whether the final demand due date schedule can be met. A schematic visualization of these first few MRP modules appears below in Figure 10.1. The basic problem in planning for final demands is that of predicting them.

DISTRIBUTION DEMANDS

If a manufacturing firm had only one product outlet and only one product it could plan its production schedule by forecasting demand or by producing only to fill firm orders. Then it could start with order due dates, explode resources needs backwards, and acquire a complete schedule of supply orders and pro-

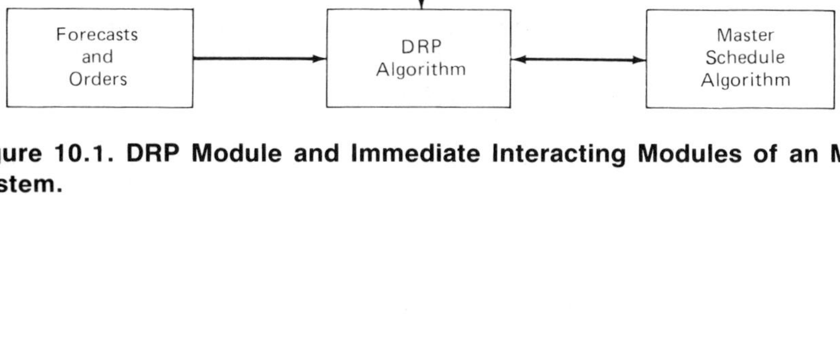

Figure 10.1. DRP Module and Immediate Interacting Modules of an MRP System.

duction runs, using verified data concerning lead times, capacities, and standards discussed earlier.

If the firm had only a few products, it could do much the same, except it would now have to assign product priorities to determine which orders to delay if production bottlenecks or delays force revisions in the master schedule. If the company made a few products and then acquired a small number of distribution points, all nearby, it would not need to complicate its planning procedures drastically, although it would probably allocate safety stocks to the individual sites, thus enlarging its investment in inventories.

However, if the number of distribution points is large the company faces the classical difficulties of distribution planning, which result from conflict between local and global forecasts and from long distribution lead times. Let us consider the sources of conflict over forecasts, caused in part by the length of lead times.

FORECAST CONFLICTS

The more distribution points there are, the farther from central production facilities they will usually be and the larger the safety stocks at each site will be. However, field managers are often nervous about their stock levels because they know they are difficult to replenish rapidly. Because of this fear, field personnel often submit forecasts much larger than market conditions justify. This forecasting ploy creates conflict with centralized forecasts, which are sure to be lower and to call for smaller safety stocks.

A strategy suggests itself. It consists of selecting some field sites to operate essentially as just-in-time locations while selecting other sites as buffer stock locations. By adroitly choosing the buffer sites to dovetail their coverage of regions to include the just-in-time sites, total safety stocks can be reduced without losing the ability to avoid stockouts in any region. A just-in-time distribution point would be stocked strictly from firm orders, while a regional warehouse would be stocked as per a regional demand forecast plus regional safety stocks.

Let us examine the nature of forecast conflicts in a bit more detail. It is not unusual for a firm that serves a large area to develop a single forecast per period up to its planning horizon for each of its major products. The criteria that motivate central planning include assuring that strategic inventory goals are met—such goals as turning inventory six times a year, a goal that suggests safety stock minimization. However, many firms do not attempt to break their macroscopic forecasts into a forecast for each distribution point.

At the level of a single distribution site, the attitude is often that of the individual salesman; he simply wants to be sure he can fill any orders that come in. Thus the local policy is motivated by reasoning much like that involved in

EOQ inventory policies—holding costs are weighted against the perceived other large cost, which in this case is the cost of lost sales. It is impossible to put a definite value on the number of sales lost by not having full stocks at all times, but at distribution sites the loss is often felt to be large. In consequence, lot sizes requested by local distribution points include sizable safety stock allowances.

When the macroscopic forecasts are compared with the sum of the field estimates, and the discrepancy in the quantity of products to be distributed is large, the problem faced by corporate headquarters is how to resolve the conflict. There are several possibilities.

If production and distribution lead time can be reduced sufficiently, all sites can be run by responding only to firm orders.

If buffer stocks must be maintained, keep them at the production sites and distribute them as necessary. Then stock all distribution points to fill firm orders only.

Use a combination of regional supply sites supplied with buffer stocks and individual sites supplied with just the forecasted quantities of products.

Use a mixture of just-in-time sites and safety stock sites to work much like the regional warehouse concept.

Develop site-by-site demand forecasts and stock the sites according to these forecasts plus safety stocks.

No matter which distribution policy is adopted, it must be selected on the basis of well-researched bottom-up analysis. Providing this type of analysis is naturally the province of the I.E.

DISTRIBUTION ANALYSIS

Predictability of operations is the goal of I.E. bottom-up analysis in all areas of manufacturing activities. In the specific instance of distribution the goal is to determine how to distribute product to distribution points to most closely satisfy strategic criteria.

In studying multiple distribution points, the investigating I.E. must try to reconcile conflicts between macro and micro demand forecasts. The first step is to assemble historical data relating stocks at the distribution points to local demands and to aggregate corporate demands. If these data are not routinely recorded in usable form, then the first step is to devise an information system that will repair this shortcoming.

Information must be timely to be useful in refining production and distribution goals. It is often the case that a corporate reporting system exists but is too cumbersome and slow for practical use. A general truism about large, corporate computer systems is that they tend to be comprehensive, correct, and useless.

In the case of distribution analysis, forecasts must be supervised as quickly as possible. Therefore, no more than a month should lapse between sales and their reporting if master schedules are to be able to be intelligently revised for the major products. For each major product the minimum data required are:

Sales during the report period
Starting inventory
Closing inventory

In addition, forecasts for each major product for each period up to the planning horizon for each distribution point are needed so the prediction process can be systematically reviewed and revised.

The strategic aims of distribution policy will affect inventory management and production planning. Inventory levels determine the ability to respond to fluctuating demand and account for a substantial part of product costs. Inventory policy also interacts with production scheduling and helps determine the master production schedule. The smoother the schedule, the lower the production costs.

Even when policy is strongly service-oriented and provides large safety stocks at distribution points, close monitoring of how well demand predictions match actual sales can help reduce total inventories. Simply knowing a recent count of total inventories aids a planner in avoiding excess production to inventory. The surprising fact is that many companies with large distribution systems have no good idea how much inventory is on hand nor do they have any way to find out except from out-of-date computer printouts.

Consider the effect of knowing recent inventory counts. Say that inventory policy allows 1,000 boxes of safety stock of gumballs at the Topeka warehouse. Sales projections through the current planning period total 4,500 boxes, and last month's closing inventory at Topeka showed 1,400 boxes. The 400 boxes in inventory can be counted toward satisfying projected demand over the planning period, permitting the production plant in Carbonville to schedule 400 boxes fewer in its manufacturing schedule. A saving of production and holding costs for 400 boxes will result.

In common practice, however, timely inventory counts do not exist, and the usual result is that not only will the extra 400 boxes be produced but even more boxes will be made to make up for ignorance of actual counts. Even the most advanced forecasting system is helpless in the face of inaccurate data. An information system that provides the correct, timely data is essential to the operation of efficient distribution strategies.

The natural question that occurs to an I.E. is "How can distribution points be supplied in an optimal fashion?" Among the various tactics described above, from permitting widely spread safety stocks to allowing only regional buffers to adhering to just-in-time deliveries, what is the "best" approach or mix of approaches? First, a characterization of "best" must be defined. Is it best to try to minimize the amount of inventory or to minimize the probability of stockouts or to minimize real costs or is there some other criterion dictated by broader strategic goals?

If an I.E. is trying to approach this question of optimal distribution policy, he must first understand the array of corporate objectives within which that policy must fit. In some companies the dominant criterion for choosing every policy is that the policy maximize capacity utilization. In such a company demand swings are of no consequence and inventory accumulations are ignored. Some firms producing aluminum products have been governed by capacity maximization strategies in recent years. An I.E. who understands this to be the policy would know not to be concerned about the reliability of market forecasts; however, most firms operate with much more concern for demands and inventories.

Usually there are tradeoffs between production scheduling, transportation scheduling, and inventory costs, and an aversion to missed sales must also be considered in seeking optimal procedures. An I.E might try to model the problem as an LP problem so he could run the LP model each month, using the current input data, to satisfy the multiple criteria. One criterion, say costs, could be minimized while other criteria, such as stockouts, could be held to acceptable figures in the model.

Another modeling approach uses a combination of statistics and simulation in which historical data are used to develop statistical profiles of local demands, which in turn are used to drive a simulation of production and distribution over the current planning period. Production and distribution policies can be revised between runs of the model until it shows acceptable average levels of decision criteria, such as number of stockout conditions or sizes of inventories. The resulting policies might not be optimal, but they would probably be quite good.

Currently, there are a number of software packages available that can do the simulation described satisfactorily. There are statistical packages that can handle complicated time series data that go into making market forecasts, also. There are, in short, many tools to aid the I.E. in bringing coherency to distribution/production/inventory policy. However, there are problems that arise from special situations, such as promotions, that must also be addressed.

PROMOTIONS

Promotions are infrequent drives to improve sales of established products or to arouse consumer interest in rising products. Both cases create exceptions to routine demand forecasting. Promotions of established products are intended to increase expected demand, while promotions of new products are part of long-term product development goals.

New product programs may not be expected to turn an immediate profit, which is the important exception about them. In addition, it is extremely difficult to project sales of a fledgling product. Given these conditions, there is little point expending effort to sharpen a new product's demand forecast, when it is only a guess to begin with.

The case of promoting an old product is quite different. Over time, consumer reaction to promotions of a given product can be monitored, and the recorded data can be used to predict the effects of promotions on demands during a promotion period.

An astute I.E. will note that acquiring such data requires that information systems exist permitting the collection of sales information. When he is installing such a system, the I.E. should be careful to analyze exactly what data he will need for predictive purposes and then design his information collection system accordingly.

As we have noted earlier, it is important to be able to predict not only corporate, but regional and local demands. Therefore, data collection should be sufficiently detailed to gather data at each level of detail. Any other items it is advisable to record for predictive purposes should also be anticipated prior to inaugurating the information systems.

Another important topic is price elasticity of demand (i.e., how large an increase in demand will changes in unit prices cause?). Past data may be sufficient to project the next promotion's effect on sales by simply fitting the old data to a trend line. It may be that, for prices near the current price/unit, each $1 cut will induce 100 more people to buy the product in each region during the promotion period. This assessment provides a functional basis for revising production and inventory levels prior to the beginning of the promotion.

The concept of planning ahead for promotions often is limited by habitual procedures. Promotions are usually instigated by the marketing department of the firm. Unfortunately, many marketing departments exist that do not talk to their production departments. An aware I.E. should take note of whether his firm's marketing and production departments converse, and, if they do not, he should try to see to it that they begin.

Especially in the case of promotions, it is important that the production department know when extra product will be required. This information needs to be made available soon enough to provide for smooth schedule changes. It is also important for sales and distribution personnel in the field to be informed of promotions soon enough to prepare. For instance, a warehouse manager might wish to change the location of storage for a promoted item, and sales managers may add temporary help during anticipated rush periods.

An I.E involved in the preparation and implementation of strategic plans should be especially alert to determine whether the implied information network with lead times short enough for information transmittal exists. If the system does not exist, he should make management aware of what kind of network there should be and why.

Just as the locations, stocking levels, and demands are important to product distribution, so is the selection of transportation routes and carriers. This topic naturally seems to require I.E. action.

TRANSPORTATION

Every I.E. has been exposed to the Transportation Algorithm in which various routes between warehouses and stores are assigned a cost representing how much it costs to transport one unit of a single product from each warehouse to each outlet. The model for the problem of choosing the least-cost mix of routes is a linear programming model which is solved by LP or by the transportation algorithm or by a branch-and-bound method.

Real-life transportation problems are much more complex. The first complication is that there are several different products to be shipped at once to each site via volume-constrained carriers. It is best if a carrier can be fully loaded, else the cost per unit for carrying the load goes up. The second complication is that a different mix of shipping modes exists for different distribution points. One site may take only barge and rail shipments; one site may only receive truck deliveries. Costs tend to vary significantly from mode to mode and from site to site.

How one arrives at the simplistic costs per unit of the classroom is hard to say. The only way this could be done is if identical shipments were made over and over; then fairly reliable cost averages could be obtained. In real life very rough cost estimates must be made, and judgment calls must be made frequently because all carriers are not always available. Strikes, mechanical breakdowns, and other causes can reduce the range of shipping options at any time. Even if a firm knows its optimal shipping strategies, it may not be able to implement the plan due to outside interference, even weather.

This is not to say that the effort to ascertain the best routes and shipping modes should not be made. The best available computational aids should be used, and the best cost estimates should be acquired. Often the disparity between alternatives is so pronounced that even the ''noise'' in the data does not confuse the decision. A second reason to conduct a transportation study is that it provides an understanding of the issues that may have been lacking.

The effort consists of assigning costs to the routes and the modes of transportation. Either volume or weight of product can be the basis for costing transportation. A box that is 8 cubic feet in volume is a fixed percentage of the volume a given carrier can handle. Similarly, a 1000-pound item represents a fixed fraction of how much weight a truck can carry. It may be that simply multiplying that fraction times the full cost of using that carrier will produce the desired cost.

It is acknowledged that this cost figure is incorrect unless only full loads of single products are shipped. If a mixture of products is loaded on a truck, the sum of the unit transportation costs for those products will probably not be the same as the actual transit cost for the truckload, but if the cost/unit numbers provide a close estimate, the responsible I.E. will be happy. Instructive results from the cost-gathering exercise are the discovery of how and why routing costs vary and the determination of how much of the cost of an end product is attributable to transportation.

The lessons learned are analogous to those learned about other corporate operations during activity analysis. It is not unusual for a company to suddenly discover that its transportation costs are high enough to make some products lose money. In some cases it will be found that only shipments above a certain size are cost-effective. The repercussions of this discovery will be felt at distribution points because it mandates a minimum lot size for shipping.

In rare cases it is found during transportation analysis that some products have transportation costs too high to allow them to ever turn a profit. This is a good fact to find. Surprisingly, many companies have no idea what unit costs and route costs are and habitually ship via the first available carrier. This is not an intelligent procedure, but it is difficult to inspire management to question a habit. Still, such questioning is exactly the job that a strategically aware I.E. should perform.

SUMMARY

Distribution plans, if they are to adhere to strategic objectives, must be put on a firm factual basis. Intelligent inventory levels must be set and well-formulated demand estimation and distribution procedures must be derived from accurate data. Marketing and production must also be coordinated if effective distribution techniques are to be implemented. Finally, transportation throughout the distribution network must be scientifically analyzed and put on a program of cost-effective planning that conforms to strategic objectives. I.E.'s are ideal personnel to develop the data base, devise information systems, coordinate the activities of diverse personnel, monitor effectiveness of distribution plans, and serve as distribution experts during strategic planning sessions.

11
MARKETING AND APPROACH TO COMPETITION

Marketing is a corporate discipline that influences both short-term and long-term aspects of strategic planning. To some degree long-term marketing policies are based on the types of markets the company is targeting. Brief outlines of some of the main types of markets follow.

YOUNG MARKET

Shortly after a market has been entered, it can be expected to be in turmoil. The home computer market is a good example of this condition. The market attracted far more producers than could survive the competition, so the expected result was that a shakeout of competition would cause some competitors to fall by the wayside. This indeed happened. Texas Instruments, Osborne, Coleco, and other companies suffered as a result.

A young market presents numerous uncertainties to market planners. Clearly defined market segments are not yet developed, so market planners are forced to guess at market forces. In the home computer market two early strategies were: (l) to price low enough that impulse buyers would not be discouraged from buying, and (2) to provide consumers a sufficiently powerful system to satisfy a serious computer user.

The first strategy limited the hardware configuration of the computer and put a de facto lid of about $200 on the price. Within this category there were numerous competitors at the start. The second marketing approach allowed more flexibility of strategy than did the low-price ploy. A variety of direct consumer incentives, such as low-cost peripherals and software, were provided in addition to attractive prices.

The most important consideration to the serious computer user was the reliability and availability of useful software. To enhance its position in this regard, IBM openly encouraged imitation of its hardware, which in turn led software developers to favor development of IBM-compatible software, all of which enhanced IBM's position in the microcomputer market. Even this strategy could not save IBM's PCjr, which was apparently priced too high for the home computer market and which was not powerful enough for the serious user.

This market is probably still very young, so the ultimate ends of developing strategies are not evident. Computers are subject to drastic changes in size, capabilities, and price due to rapid technological changes, so much so that computers may retain young market characteristics for many years.

MATURE MARKETS

The lessons to be learned from the computer experience is that young markets are unpredictable and may require rapidly changing strategies, or at least rapidly shifting tactics. Thus young markets are difficult to address in long-term strategies. General advice on how to approach young markets is also difficult to provide. Unless product interest dies, a young market eventually matures into a mature market, and general advice does apply then.

In a mature market there are few surprises. The market niches have all been staked out. Price ranges are fully understood, along with the nature of the consumer. Few strategies exist for increasing market share in mature markets. Cost-based strategies, involving cutting prices by learning how to cut costs, may be available, but cost cutting when procedures are well established is not easy.

In the absence of outside forces that cause a market to restructure, businesses in mature markets tend to be much alike. They use the same production layouts, the same marketing approaches, and much the same management techniques. A manager can go from one competitor to another with little retraining. High-cost approaches have shaken out over time, and competitors have become virtual clones.

Consider the U.S. automobile industry prior to 1973. Cheap gasoline and meager public interest in air pollution had permitted all the major car makers to adopt essentially the same approaches to the marketplace. They all turned out large cars with high profit margins, cars that were usually overpowered and underdesigned. The sudden high price of gasoline, coupled with sudden government intervention, forced the market to change drastically. Without the influence of forces for change, it is hard to see how the Japanese could have become serious competitors in the U.S. car market.

A recent version of conventional wisdom holds that a company should not allow its business to become dominated by mature markets. One reason for this view is that mature markets have a way of becoming saturated.

SATURATION

Imagine the home computer market in several years. It is easy to see a situation in which everyone who wants a home computer and can afford one has one; the market will be saturated. This may have already occurred. Any company that insists on doggedly making products for a saturated market must count on sales to new entrants into the market plus replacement sales. However, the manufacturing capacity required to sustain mature market sales is ordinarily smaller than that required to take that market through its growth period. Unless this condition is anticipated and the excess capacity is planned to be redirected, it can be a severe drain on profitability.

A common recent approach to dealing with the effects of market saturation on a company's growth is its diversification into markets that show more potential. Acquisitions have been the usual mode of this kind of diversification. Diversification strategy is often aimed at developing a market mix that balances the pros and cons of its segments. Thus, a company in a mature market might very well acquire a company in a young, volatile market, the acquisition being the corporate equivalent of adding a speculative component to its investment portfolio.

Other market factors that must be considered in developing strategies are how long the products last and how long the markets last. Estimates of these periods allow the prediction of when markets will become saturated and when they will die altogether.

PRODUCT AND MARKET LIVES

Almost every product has a life expectancy—either planned or random. A planned life cycle [52] might be aimed at a mature market such as razor blades or telephones to make a slightly different product seem worth buying. A product with a planned death will be withdrawn and replaced with its next version when the appointed time arrives. Unplanned life cycles can be either long or short. Brick has been a popular building material since the time of the pharoahs, so its life cycle is quite long. Hula-hoops had a life cycle of almost infinitesimal length.

Many electronics devices have a very short life. For instance, new circuit chip capacities and materials are continually being developed, leading to very short product life cycles. Here a distinction must be made between a product's useful life and its life cycle, i.e., the period during which it is in demand. An out-of-date circuit board may last 100 years as a workable part of a device, but its life cycle may be no more than 6 months.

The tradeoff between these two time periods is important to product planning. Consider the color television market. If this market is not now saturated, it is close to it. In addition, the underlying technology has become steadily more reliable until repairs and replacements now are not nearly as frequent as they once were. The longer the life of the product, the smaller the replacement rate. Thus, even though color televisions appear to have a long life cycle, they have a small need for production capacity.

Many products have a short life cycle. If a company believes its product to have a short life cycle, it should plan well ahead to have a new product to take its place as a contributor to company cash flow. A good example of this situation might be CB radios. Although they were not a new product, they were not a mass market item until they suddenly caught the public fancy. Any company that jumped into CB production in the middle of the surge should have had a backup product on the drawing board if it were to be able to sustain capacity utilization, because CB demand died rapidly.

Companies must be able to anticipate sales if they are to plan. Forecasts of slow-moving, long-lived products with enduring demand must be made, as must estimates of the life cycles of more mercurial products. Product introductions should be timed to maintain a consistent level of sales despite rising and falling demands. The goal is to assure stable total demand. Toward this end, manufacturers must understand the cyclic nature of products.

CYCLIC AND SEASONAL VARIATION

The long-term variation in demand resulting from the life cycle effect must be considered in drawing up strategic plans, as must the shorter-term effects of cyclic and seasonal demands. Seasonal fluctuations commonly happen to demands for products that are heavily requested by consumers only during one season of the year. Department stores base their profits largely on Christmas sales, for example. Cyclic effects are more difficult than seasonal to predict; they are the fluctuations over periods as long as several years or even several decades.

The oil business now appears to be in a long demand trough, but most analysts feel that conditions of low supplies and high demands will recur eventually. Anticipating the rising demand would be a significant activity at a petroleum-dependent enterprise. It should be noted that there is no certainty that oil will be a future cyclical business; only past data can be definitely labeled as cyclic.

Some very long demand changes that are fairly predictable if not cyclic are those that are strongly related to demographics. The age mix of a population can be predicted many years forward, given the current birth and death rates and age mix. Baby food companies eye birth rates very carefully, as do all companies that target sales to specific age groups.

Once a firm has a good understanding of predictable demand fluctuations in markets, it can intelligently plan programs of market development with reasonable strategic marketing goals. These goals might be leveling profits, reaching growth targets, or developing into new markets altogether.

STRATEGIC ASPECTS OF MARKETING

Strategy must start with knowledge. A simple-minded plan to "increase sales by 50 percent" will appeal superficially to many managers, but such a proposal must be achievable to be meaningful in strategic plans. A range of plans may all be possible. For instance, a department store may be content to rely on its annual Christmas sales for its existence. Many stores take this approach.

It is more common for manufacturing firms to try to remove uncertainties from the demand picture. In cyclic industries a common strategy has been to diversify into noncyclic product lines or to develop countercyclical products. Another approach has been to develop service functions that complement the primary business. A good example is an equipment manufacturer that also employs its expert personnel as manufacturing systems consultants.

As in the case of distribution planning, market planning can be approached by use of categorization. First, the company should identify which portions of its business over the strategic planning period will be fairly stable. This analysis will consist of studying past data and performing in-depth market research. The role of an I.E. in this analysis would probably be that of a coordinator of the efforts involved.

Secondly, seasonal and cyclic demands must be anticipated. The analysis needed consists of data review and market research. Lastly, short-lived products, those expected to have short life cycles, should be projected. This effort will require cooperative work between product development and research personnel and market research people, perhaps under the coordination of an I.E.

The less reliable the forecast of demand, the more risky the enterprise. Thus, short-lived products are speculative, but mature products are not. Cyclic and seasonal products have risk factors somewhere between short-lived and mature products.

Strategic objectives can now be defined in terms of the three demand and risk classes. One immediate strategy is to minimize risk. This suggests introducing speculative products or attempting to penetrate uncertain markets only as a last resort. Another common goal of strategy is to maintain profit margins or total profits or ROI. This strategy may be more difficult to implement than minimizing risk because it requires constant effort to create new market niches to counter sales attrition due to competition or saturation. An even more difficult objective is maintaining constant growth, which implies steady development of new products. It also creates the need for steady new investments of capital to increase productive capacity. Larger capital costs create strategic imperatives of their own, such as the need to maximize throughput in new facilities in order to reach breakeven points made higher by capital-related costs.

The riskiest strategy of all is diversification for the purpose of minimizing the effect of the failure of any one project. This is the strategy that has created today's huge conglomerates and that has produced the recent "acquisition mentality" among managers. Rampant diversification, even if it is temporarily effective, is counterproductive in the long run. Fully diversified conglomerates find themselves faced with the same situation that faced mutual funds a decade ago—once a stock portfolio is fully diversified, that portfolio imitates the stock market as a whole almost exactly. Analogously, a fully diversified conglomerate will imitate the economy as a whole, including its cyclic behavior.

Some companies are more inclined toward accepting risk than others. This proclivity almost seems an inherent trait and is often considered as part of the corporate culture. An I.E. trying to understand his company's strategic plans first must ascertain what its unwritten beliefs are. Some large firms even publish their ideologies, although it is not certain that their recorded beliefs are those that are acted upon. For instance, IBM espouses such aims as product reliability, employee development, and customer service, but its actions reveal its dominant strategic impulse is to bury the opposition.

Once a strategically aware I.E. has objectively determined his company's natural impulses, he is in a position to test whether they conflict with the strategic plans that are nominally in force. If the two do not correspond, they will surely conflict. Mixed messages will be spread throughout the firm. For instance, many an engineering school gives lip service to a desire to do quality research whereas the true motivation is a desire for research dollars without regard to quality.

The I.E. has a problem if stated and real goals do not correspond. He is put in the position of trying to make the strategic plans somehow more rational. This effort may require substantially changing the corporate culture, a notoriously difficult enterprise. When managers are told realities contradicting their predilections, they often rationalize the facts and maintain their preferred views.

If this happens to an I.E.'s presentation of facts, he will be frustrated in his attempt to revise unrealistic strategies. He must then decide whether the continued frustration is worth the occasional success or whether he should seek employment in a firm whose philosophy more closely fits his own. A disreputable, but common, third route is to resign oneself to performing routine duties in an atmosphere of illogical plans.

Once an I.E. has understood the company's strategic agenda, he can perform the bottom-up analysis needed to refine the plans into practical procedures. Much of the detailed analysis required has already been discussed.

OPERATIONAL ASPECTS OF MARKET PLANS

The simplest part of the production plan descending from a market plan is that pertaining to routine work associated with mature markets with flat demand. Data from activity analysis and capacity analysis make this portion of the production activities easy to plan. All operations from distribution to inventory control to production to supply sourcing should be routine.

Cyclic markets are more difficult to anticipate, and they seem to require special procedures. One approach to eventually satisfying unknown demand with uncertain timing is to constantly maintain large safety stocks. This can be accomplished by producing to inventory during light periods in the production schedule. This strategy suggests that cyclic products may require a different inventory policy than do flat-demand products. If this is in fact the case, then the special cyclical inventory policy must be tested against strategic goals and profitability criteria to see whether it is permissible. Two other approaches to cyclic demand suggest themselves.

The most common approach adopted is to assume that, when the demand begins the upward part of its cycle, production capacity is sufficiently flexible to accommodate the increase. Lastly, a portion of the plant may be reserved as buffer capacity, infrequently used but available for cyclic products without affecting the part of the plant dedicated to routine activities. However, completely separate buffer facilities would be difficult to operate effectively, since they would require a cyclic labor force, which in turn would suggest frequent layoffs, resulting in disgruntled and inefficient workers.

The products expected to have short life cycles require yet another viewpoint. It is largely these kinds of products that have inspired the buzzwords "flexible manufacturing" [67,40], although the term is also applicable to facilities that specialize in short runs of special orders. A flexible manufacturing facility is capable of rapidly designing a new product and producing it economically in short runs. This capability is necessary to a firm that has elected to allow short-lived products to play a major role in its strategic plan.

Flexibility can be acquired by setting up a completely self-contained unit, such as a new plant or a plant-within-a-plant [69], or by building flexibility into a routine production facility. Flexible manufacturing can only be profitable if items can be priced high enough to offset the high fixed costs and setup costs involved. Thus, it applies to unusual products that are infrequently ordered and to products expected to be not too short-lived to benefit from either economies of scale, experience curve effects, or the accumulation of product inventory.

The electronics industry has presented many examples of companies overspending capital to chase rapidly changing product markets. The cost of pursuing these markets can be more readily absorbed by a larger firm that does not rely solely on this portion of its business, but it is philosophically important for most businesses to develop some speculative products.

One reason for this fact is that people need stimulation— excitement—in their jobs. They need to feel pride in their company. Excessively conservative strategies are boring. In addition, the fallout from state-of-the-art endeavors is often materially rewarding. The work of product brainstorming, research, and development often leads to product refinements, production improvements, and unanticipated new products. The speculative part of the business is where the adventure lies, and the better personnel need challenge to feel fulfilled.

Some market strategies commonly adopted are based on technical procedures that I.E.'s should learn to understand. Day-to-day implementation of these procedures requires that supporting information systems exist and function properly, and the responsibility for supervising these systems may rest with corporate I.E.'s.

ANALYTICAL PROCEDURES

Product positioning often requires that prices be set so low that they exert pressure on the production process to hold its costs down. This situation can occur during either market development or product promotion.

One tactic used to gain early market share in a new market and one that has been used successfully is that of selling at a loss. The ultimate success of this ploy depends on the company's ability eventually to drive its cost per unit below the breakeven point while increasing its market share. This approach is based on a theory originally applied to entire industries. Theoretically, cost per unit (in real dollars) should decline over an entire industry as production experience accumulates. The curve that describes this effect is called the "experience curve" [24,38,73]. Recently experience curve theory has been applied at the individual company level.

In this application a company will plan its product positioning strategy around targeted costs per unit. In the individual company the forces that drive costs per unit down are the same forces that do so in entire industries—displacement of higher-cost procedures by lower. Say that the implementation of a company's strategic development plan for a specific product requires the cost per unit to decline 10 percent for each doubling of cumulative production units. All promotion and pricing activities will be geared to this cost reduction schedule.

The question is how will the schedule be met? It seems reasonable to expect that I.E.'s would be involved in monitoring and directing the required cost reduction efforts. Only fully allocated costs, including the cost of assigning an I.E. to the project, should be used to measure cost per unit of production. The search for reductions should center on the major cost components—labor, materials and supplies, and capital-related costs.

Labor costs are known to decline slightly in assembly line work because specialized workers acquire steadily better facility for a time at their specialties. This is not a long-lived effect, but it has a broader corporate counterpart that can help tend costs downward over a longer period. As management acquires greater knowledge of a project, e.g., making and promoting a new product, specialized tasks and personnel assignments required for startup become needed less. Thus, over time, fewer personnel and resources should be required to keep newer ventures on track.

As product costs are tracked on a basis of fully allocated costs, then as fewer hours are devoted to a new product, the labor component of its cost per unit will decline. Predictions of personnel reductions should be left in the hands of a trained professional, probably an I.E. Manpower assignments require determining when specific startup procedures have been completed and when nonroutine activities associated with product promotion become sufficiently routine to assign them to lower-paid personnel. Fewer and fewer personnel hours, especially at the staff level, should be required as a product catches on, and, finally, the I.E. can take himself off the case.

New products always create new problems. Confusion at the level of production and distribution is naturally generated by new activities. It detracts from accuracy of plans and causes costs to be higher per unit at startup than after procedures stabilize. Once the bugs in production and distribution have shaken out, the real I.E. work of cost reduction can begin, work with methods and materials.

A good I.E. assumes that nothing is done as well as it can be. Starting from this mindset, he will examine methods, flows, materials, schedules, and inventories for efficiency and logic. Lowered scrap, smaller inventories, smaller labor requirements, shorter lead times, smaller rework loads, and fewer systematic errors will be his goals.

As with all other procedures discussed in this book so far, the attempt to plan costs from the experience curve requires an information system that can produce the numbers needed to instigate improvements. Knowing ahead of time the kinds of improvements he anticipates should guide the adept I.E. as he installs information tracking procedures. It costs money to obtain and analyze data, so the cost control plan should include savings from the steady reduction in complexity of the new-product data-tracking system. It should eventually be absorbed into the company's routine data system.

New products present a strong case for the use of fully allocated costs, including capital costs. If the company's facilities are unaffected by a new product's introduction, then one could argue that the product's capital-related costs could not decline. However, if costs are fully allocated, machine costs should be charged to a product according to its machine requirements standard. Thus, as a new product comes to be made more quickly, the cost of machine time should decline per unit.

In summary, the basic requirement for using experience curve principles in casting product plans are: (1) a fully allocated cost system must be in effect; and (2) personnel, such as I.E.'s, who are cost-conscious, production-knowledgeable, and strategically aware, must be assigned to direct and monitor the cost reduction effort.

Roughly the same set of business consultants who promulgated the experience curve concepts have recently added a frequently cited idea to conventional wisdom. The idea is that some portions of a company's business should be viewed as cash cows, i.e., steady but unspectacular money producers to be milked for cash to invest in more promising parts of the business, especially speculative ventures. This idea seems superficially appealing, but in practice it can easily be carried to an extreme.

Say a company has a stable share of a mature market. If this market is viewed as a cash cow, the company should invest little or no money in this segment of its activities and allow it to slowly die. However, a company needs some dependable activities as a base from which to support cyclic and speculative activities. Also, there is no good reason to expect that every stable market is a dead end. If this were the case, makers of refrigerators would have all gone out of business years ago. Thus, a reasonable approach to many cash cows that show fairly stable profitability is to sustain them.

In fact, a stable business may periodically offer opportunities for large cost reductions due to newly developed technologies or materials. It is true that some facilities and some product lines outlive their ability to make a return on an investment, but these cases must be decided case by case. Reductio ad absurdum, the milking of all cash cows to feed speculation would eventually make a company totally speculative. Few companies could survive in that mode.

SUMMARY

As with all strategy development, market development plans must be drawn with specific goals derived from accurate intelligence. Data needs include market forecasts, profiles of markets, accurate assessments of the company's profits and costs associated with existing product lines, and a full understanding of product demands. Profitability criteria should lead to a strategic mix including both low-risk and speculative products developed from a long-term market development plan. All information systems needed for implementing and monitoring the plan must be in place prior to initiating its actions. Trained personnel, such as I.E.'s, who fully understand market, strategic, and operational concepts must be assigned to keep the plan on track.

12
ORGANIZATIONAL STRUCTURE

Humans instinctively create two kinds of organizational structure—vertical categories of hierarchy and horizontal divisions for specialization. The vertical structure is determined by ascending levels of authority and responsibility. The specialized skill necessary to perform many corporate functions creates discrimination between functions and leads to their being assigned discrete categories on organization charts.

All human endeavors tend to proliferate classifications, so it is natural for organizations to constantly add more hierarchical levels and more job classifications. The fact that this is a natural occurrence leads corporate personnel to unconsciously accept burgeoning organization charts and expanding payrolls without question. One role an I.E. may find for himself is assisting to make organizational structure more rational. However, this can only be accomplished when the subject becomes one of conscious attention. Only then can it be discussed rationally and included in strategic considerations.

STRUCTURE AND STRATEGY

A common cause for a company not meeting strategic objectives is its inability to implement them rapidly enough. In the case of market development or product introduction, hesitation can lose market share. The last entrant into a market may be unable to overcome early market development by competitors. There are many possible causes for late corporate decisions and action, but a common one is unwieldy corporate organization.

Slow decisions and late implementation of plans often occur when there are too many people involved in the decision making. It has been estimated that some U.S. corporations have had as many as 14 levels of decision making. If each decision were to require a week to make at each level, a decision would take 14 weeks to work its way through the heirarchy. Such a pace is far too slow for competitive responses in volatile markets.

Another source of slow decisions is inherent in an approach to decision making that many Japanese corporations use—consensus management. U.S. businessmen dealing with Japanese firms often find themselves involved in seemingly endless meetings that make no noticeable progress. The Japanese seem happy to postpone any decision until everyone concurs.

A company afflicted with too many decision levels probably has too many levels involved in implementation as well. Even if a strategically sound decision has been reached, translating the decision into action may require the gradual descent of actions through each of the decison levels as actions backtrack the path of decisions. Consensus management does offer the advantage that, once a decision is reached, the enactment of the plan is almost immediate.

There are advantages and disadvantages to each of the various approaches to decision making and decision implementation. Often the most efficient procedure is to restrict decision authority to the fewest personnel possible. Sometimes the decisions hold far too much strategic significance not to be subjected to the scrutiny of several decision levels. In U.S. companies consensus is rarely a practical decision procedure except in small groups.

Efficiency and cost-effectiveness should govern the choice of decision making and implementing procedures, and the resulting organizational structure should conform to the company's strategic objectives. Not only must horizontal and vertical structure be established; specific lines of responsibility must be drawn, and information systems to facilitate plans at all levels must be installed.

It is widely argued by organizational researchers that decisions should be made as close to the level at which they must be implemented as possible. This implies that a plant manager should make basic production decisions and that a marketing manager should have final authority in dealing with his salesmen. There are several good reasons for adopting this policy as part of the corporate long-term strategic plan.

First, the manager closest to the problem usually has the most knowledge about how to deal with it. Second, the closer to the problem the decision is made, the quicker action is taken, and speed of action is often critical. Third, the more responsibility a good manager is given, the better he reacts, and the better he does his job. Drawbacks to giving authority at a low managerial level include the possibility of low-level personnel being inferior and the possibility that the low-level decisions may have strategic implications requiring higher-level scrutiny.

As has been persistently stressed throughout this book, decisions made at too high a level may be made in an atmosphere of ignorance. Thus it behooves higher-level management to seek the assistance of an interlocutor between strategic and operational considerations when the higher managers consider plans impacting the operational level. An excellent choice for such an interlocutor is a strategically aware industrial engineer.

But this is only one role for an I.E. relative to organizational structure. A more fundamental role is that of studying the structure, determining the real needs of that structure, suggesting improvements, and monitoring the activities of organizational segments. Surprisingly, critical organizational issues, such as white collar productivity, have received little quantitative scrutiny. Yet this is the kind of attention that an I.E. can bring to these issues.

Organizational structure tends to grow; new job and hierarchy classifications proliferate. The most dangerous aspect of this proliferation is that it is difficult to reverse; the natural instinct of many managers is to create more employee positions. C. N. Parkinson [23], in his famous study of organizations, noted that managers naturally want to increase the domains over which they rule. This desire leads to the creation of both additional decision levels in the hierarchy and additional specialties in the horizontal structure. This is one source of structural proliferation an investigative I.E. should expect.

Individual employees can also create proliferation. A person who creates a new specialty also creates job insurance, since no one else knows exactly what the new specialty consists of. The implied job security is a strong personal motive for creating a new job definition. A good example of a newly created discipline is that of Strategic Planning Manager.

Theoretically, a Strategic Planning Manager directs the company's strategic plans. It is difficult to say how such a position can be justified logically, but several reasons for the existence of this specialty can be seen. Many managers feel that strategic planning is a good thing but do not want to do it personally and hire someone else for the job. Other managers are vaguely aware that strategic planning is supposed to be a good thing, feel guilty about not doing it, and appoint a Strategic Planning Manager to assuage their guilt.

However, the only person who can be the director of strategic planning in a company is someone at the top, such as the C.E.O., the president, or the chairman of the board; it depends on the exact lines of authority in the individual firm as to which person it should be. Other individuals may have a hand in advising the leader or in making strategic suggestions, but the reins of strategic planning and control must be in the hands of the top man, the reigning executive. A management position that purports to take over a significant portion of the task is superfluous.

Nevertheless, the top man (assume him to be the C.E.O.) cannot allow his attention to be diverted to topics of minor importance, so he must delegate responsibilities in many areas. He should not be involved in such trivia as making minor design decisions that properly reside with product-development engineers. On the other hand, the C.E.O. cannot be wholly ignorant of any issue handled below him. Not only must he delegate authority and responsibility, he must also obtain summarized information on all important corporate activities to keep abreast of significant events.

Similarly, at every level and every category of specialization, personnel need to be fully briefed on the aspects of the business they need to understand. It is somewhat easier to define just what information a Materials Manager needs than it is what the C.E.O. needs, but the information needs of each job should be based on the nature of the job. This fact should guide the I.E. analyzing corporate structure.

If it is unclear what information an employee needs to do his job successfully, then that job must not be well-defined, and if a job is not well-defined, it may well not be necessary. Job definitions should direct the structure of the firm's information systems; they should provide the basis for the actions that ordinarily need to be taken.

Let us test the definition of the position of Strategic Planning Manager. What tasks should the Strategic Planning Manager perform? None, the tasks of formulating and implementing strategic plans constitute the main work of the c.e.o. What information needs does the Strategic Planning Manager have if he has no tasks? Obviously, none.

Similar logic can be applied throughout an organization to locate superfluous levels of management, superfluous specialties, and redundant personnel. During the boom period of the early 1970's many companies added numerous unneeded employees at the staff level to their payrolls. Some analysts refer to the result of this tendency to overhire as "corporate obesity." Staff hiring detracts from a firm's ability to afford important line employees, which in turn weakens their ability to be competitive [62].

During inflationary periods, such as the 70's, prices could be escalated fast enough to support the exorbitant costs of inflated payrolls, but when a recession reduces product demand suddenly, as in the late 70's, it becomes apparent that many more people are on the payroll than are justified. Thus, in the 80's many companies have been reducing staff sizes, though not as rapidly as logic would demand. Logic is difficult to enforce when it requires depriving someone of his livelihood. The better course would have been to never have hired unneeded personnel originally.

The question that must be answered is how can real staffing needs be estimated. The starting point for finding the answer is the strategic plan. This plan should outline operating levels and specific planned projects over a period of several years. If full activity analyses have been performed so that fully allocated resource needs can be assigned to the projected activities, then the plan's resource requirements should be estimable. When these requirements are compared with available capacities, it should be possible to assign the staffing levels needed to facilitate the plan.

However, the resource standards used in activity analysis are actually based on "how things are done now." Periodically, the wisdom of existing procedures should be questioned. Probably under the supervision of an I.E., a structural analysis should be conducted. Structural analysis of a firm seeks to relate existing functions to the personnel performing them to determine whether the authority and responsibility structure is logical, whether resource allocations to the various functions are sufficient or excessive, to define jobs more tightly, and to develop measures of effectiveness for managers, departments, and employees.

Structural analysis also relies upon flow analysis and decision structure analysis, which are discussed below, but as an independent activity it questions the need for every function in the firm. It attempts to define jobs and determine real staffing needs. It also tries to match existing information systems to necessary functions and to ascertain improvements in those information systems needed to make people more effective. Ultimately, every activity must satisfy cost criteria to justify its existence, but structural analysis tries to discover what jobs and resource commitments are necessary and how they should coordinate if the firm is to achieve its strategic goals.

Consider, for example, order entry and release, a necessary functional area of a manufacturing firm. An I.E. studying this function would first need to define the specific activities assigned to it. As an example, let us consider an order entry and release function at a single plant site of a large corporation. Assume that orders come to the plant site from corporate headquarters. The orders may be developed from forecasts or from firm orders or from some more complicated approach. Due dates and destinations along with a complete list of ordered items would constitute an order.

The plant's order entry/release office would enter each order into its order records and into the plant's master scheduling system, which may or may not be computerized. Upon completion of production scheduling, the date the order is actually expected to be ready for shipment should be submitted to the order entry/release office, so it can update its records and inform the corporate office of any due date changes.

This office should also be the proper one to track orders and determine whether they are initiated on time and whether they are on schedule. Order Entry and Release should also be informed as soon as the order is complete and ready for shipment, at which time Traffic should be given the order release for shipment. When Traffic returns the completed order release form to the entry/release office, that office should promptly update its order completion files and transmit the same information to the corporate office.

An I.E. studying the local order entry and release procedures would be interested in a number of questions about this specialized activity. He would want to know how it fits into the plant's hierarchy, whether its function is necessary, what resource allocations should be used as its job standards, how the order entry/release information systems work and whether they are adequate, what staffing levels the office should be given, and whether it works in a cost-effective fashion.

It seems clear that this function is necessary because someone needs to continuously monitor the progress of orders from the time they are received until they are shipped. Given a sufficiently powerful computerized information system it might be that only a very limited set of personnel would have to be involved in tracking orders, perhaps only the plant manager, perhaps only he and the scheduler. However, it is usually wise to have at least one person who is responsible for the accuracy of order data if for no other reason than that automatically recorded data is often wrong and needs to be corrected.

If the existing information system is not sufficiently automatic to replace the order entry and release functions outlined above, then the job will require at least the partial attention of one person. An estimate of the staffing level needed can be obtained by determining the average number of each of the clerical functions during a month and then assigning time standards for each major function. Only necessary functions need be studied; unnecessary functions should be identified and eliminated.

A study of the real information needs associated with order entry/release must be performed to discover which pieces of information are required to acomplish the office's duties. An accurate record of sales must be maintained if accurate accounting data are to be kept and if budgets and costs are to be able to be reviewed, so order entry/release must perform the necessary function of reporting shipments that represent final sales.

Just as certainly, no production facility can operate intelligently if it has no accurate record of what it is supposed to make nor when it is supposed to have made it. Thus, order entry must be maintained with superlative accuracy. Herein lies one of the reasons why so many specialized functions exist within a company— someone must feel responsibility and accountability for each important function or the function may not be done properly.

If order entry/release is manned by plant personnel whose main responsibilities lie elsewhere, then order data errors will not hold adequate significance for them. Their resulting lack of concern for the accuracy of this function may in turn interfere with the company's achievement of its strategic goals. However, if a single individual is assigned the task of maintaining the entry/release function and given the authority to do so, then doing the job right will be foremost in the mind of that individual. This line of reasoning is responsible for the creation of a new position, possibly referred to as the Manager of Plant Order Entry and Release, a staff position accountable to the plant manager.

Not only must order ins and outs be recorded with accuracy, continuous monitoring of jobs must be performed to ascertain whether jobs are on schedule. Recent management fads have invested computer programs with much of the responsibility for tracking jobs. The computer programs are excellent adjuncts to human effort, but they do not eliminate the need for someone to regularly check the data upon which the programs depend. Only a human who feels responsibility for this segment of plant activities can provide the detailed attention that constant monitoring requires. If order entry/release is an important function, and it is, then it should be assigned a functional position in the plant organization chart along with the responsibility for data accuracy and the monitoring of job schedules.

An I.E. studying this single plant function will also note that it has considerable importance at the corporate level. As was mentioned previously, market demand forecasts require accurate and timely information about actual sales, projected demands, firm order dates, and inventories. If a plant order entry/release office is very slow or inaccurate in its reporting of job schedules and order shipments, then corporate forecasts could be impaired, distribution planning could be thrown off, and inventories could be either too large or too small. Accurate order entry/release data are essential to effective corporate planning.

Each important building block of a company's information and operational structure should be studied just as order entry/release should be. In this way it can be determined which functions are essential and which are not. This study of jobs, especially at the staff level, can lead to the curtailment of functions that had previously been heavily staffed.

Ordinarily, corporate inertia would tend to keep staff positions in place and unquestioned forever, but the declining profits of recent recessionary periods have caused management to search for ways that significant cost savings could be made. Elementary analysis of functions and information flows have shown that certain staff functions are required infrequently. Many larger companies have reduced their staff sizes markedly. For example, U.S. Steel has even eliminated many staff engineering positions on grounds that the essential work can be farmed out at lower cost.

The problem that occurs in large corporations and that is exemplified by the case of the engineering staff is that unnecessary or excessively costly functions will continue being staffed unless they are specifically sought out. One way to assure that they will be investigated is to assign an objective observer with a systems viewpoint and a strong interest in cost reduction to the job. It would appear that this work of structural analysis is a job for an I.E.

The starting point for a corporate-wide structural analysis is the determination of the real functional needs and real information needs throughout the firm. Most line functions are clearly necessary, such as production. Support functions at the plant level are also readily justified, such as inventory control, materials management, personnel management, scheduling, maintenance, quality control, etc.

Higher up the corporate ladder, coordination functions are necessary: these functions include corporate planning, marketing, distribution, and inventory control. From the bottom level of activity on up, the truly necessary line functions are readily ascertainable. Genuinely necessary functions are those without which line activities could not be performed. Few people would question the validity of many corporate functions; disagreements arise largely over the need for support functions.

Examples include advertising, maintenance, transportation, and legal services. In many cases these functions can be given to outside firms to perform at less cost. Some of these support functions are needed so rarely that they can virtually be ignored. An investigative I.E. is sure to feel that a Pareto rule must apply to corporate functions—something like: "Twenty percent of the functional classifications perform eighty percent of the work."

After a first cut at identifying necessary functions has been made, the I.E. doing the analysis can further refine his list of structural needs by trying to match real information needs to functions. For example, he may find that the market forecast effectively drives the entire corporate mechanism. Once the market forecast for the current planning horizon is prepared, then that forecast is fed into the DRP calculation module, which interacts with the master schedule and the inventory control calculation procedures to produce a revised master schedule, which in turn motivates the MRP algorithm, which directs the need for supply orders and production setups. Derived from the above information will be production schedules and cost estimates that can be tracked to determine variances.

The data needed at each stage of the operations described above become readily apparent upon study, but the roles of functions that support operations are more difficult to define. However, once a structural analysis has been completed that matches informational and operational needs to organizational entities, time standards should be derived from average work loads for use in justifying staffing levels. One other step needed to refine the derived structure is decision structure analysis.

DECISION STRUCTURE ANALYSIS

Informational and operational needs will not completely dictate structural staffing needs because there are also coordinating, monitoring, and managing requirements that will demand personnel. To ascertain how many people must be in place to handle these functions, a study of the structure of the corporate decision process must be made.

The strategic plan itself, its formulation and implementation, dictate part of the decision structure. For instance, since marketing and production activities must be coordinated to effect strategic marketing plans, there must be someone to monitor and ensure the existence of this cooperation; it will not happen spontaneously. If this someone is to have the influence to make cooperation happen, he must be of a corporate rank high enough to command the attention of marketing and production managers. These considerations effectively define a management position.

If the production structure's hierarchy consists of a foreman-plant manager-production manager sequence, then the coordinating individual will probably need a rank of at least vice president in the corporation. In a smaller company the overseer of marketing/production cooperation will probably be the owner of the firm or its C.E.O. The larger the company the more duties the C.E.O. must delegate.

The implied structure consists of five decision levels and relegates most decisions to the lowest level possible. Five levels makes a lean organization. If a corporation has extensive product lines or divisions it is possible to justify one or two more decision levels, but no more. The only justification for a hierarchical level is that it performs a function, such as coordination and supervision. Monitoring of activities is a function that can be delegated and thus does not require a level in the hierarchy.

In fact an I.E., who effectively serves as an internal consultant, can perform monitoring and coordination functions that require mere cooperation rather than command. This same individual can also monitor the performance of supporting information systems to assure that they perform as required. There are ordinarily two causes for poor performance of an information system—either it is intrinsically intractable or the personnel assigned to its operation do not perform properly.

In these times when computers and their applications have taken on such a faddish popularity, many firms have purchased or leased huge information systems to govern the information flows attendant to any and all operations. Unfortunately, it is rare that a computer system will work "off the shelf" in a given corporate environment. This is a topic developed in more detail later in this book, but this fact may be the source of poor performance. The software should either be modified or junked.

If information flows are balky due to personnel problems, the usual cause is poor organization, usually too many people. Staffing proliferation often occurs because management allows too many job specialties. This problem is compounded when specialties are allowed to become superfluous decision levels, as in the case of inserting a Manager of Market Development above an existing Manager of Marketing. But the more usual cause of overstaffing is horizontal expansion in staff level positions.

Staff positions, as has been mentioned, are very difficult to hold responsible for anything, which is why it was suggested that staff time be somehow tied to line activities through indirect labor standards. However, an I.E. attempting to develop such standards may find his efforts thwarted, resulting in a lack of standards against which the staff time consumed can be compared with the time actually required by productive activities. If standards are inaccessible, one way to examine the corporation for obesity is to conduct a flow analysis.

FLOW ANALYSIS

Even a questionably staffed office will produce some records and flows of information that can be tracked, something of the sort that would interest upper management. As a case in point, consider the instance of the corporate officer who wondered why personnel insurance claims took so long to process that employees were complaining bitterly to him about the delays.

He asked the corporate I.E. department to investigate. The I.E.'s systematically tracked the paper progress of a number of claims through the personnel offices and found the bottleneck to be an office called Personnel Services. A claim would enter that office and languish for as long as two weeks, even though the average paperwork required was less than an hour per claim. A load projection over a period with an average workload showed the office could be comfortably staffed with one-half of a person instead of the five it contained. Corporate action on the I.E. report was mixed. The superfluous office was eliminated, but the office staff was reassigned, thus maintaining the same total number of personnel.

This anecdote shows that even the best I.E. work can be subverted by management, but when management is inclined to act it is sometimes possible to gain its ear and enact logical solutions to revealed problems. An assertive I.E. must understand that he will not always find his intelligent suggestions accepted. He must play a bit of the strategist's game himself and wait to strike when the chances for success are high.

SUMMARY

Structural analysis has its best chance of affecting corporate staffing policies prior to the hirings that derive from the policies. After an organization has developed habits it is very hard to make substantial changes in them. The I.E. trying to rationalize corporate structure and make it better able to implement corporate strategy will find management more sympathetic to belt-tightening suggestions directed toward departments hierarchically far from the top. Thus line workers and salesmen find a great deal of effort directed toward discovering how they do their work while staff workers responsible only to the C.E.O. find themselves scrutinized very little.

The I.E. working on problems of staffing and information flows will find his successes, therefore, to be more evolutionary than revolutionary, so he must have patience. He should not hesitate to state how and why things should be;

that is his job. And if he is able to give structural reasons for his staffing suggestions based upon decision needs, standards, information requirements, costs, flows, and corporate strategy he will eventually make an impression on even the most habit-bound managers.

The concepts of organizational analysis discussed in this chapter are extensions of ideas found in Skinner's proposals concerning manufacturing audits [32]. It must also be noted that the broad organizational perspective of this chapter is not now commonly held by industrial engineers. The unfortunate fact is that many I.E.'s have been relegated to a limited domain of activity, such as the setting and maintaining of job standards. Worse, many I.E.'s have complacently accepted this role. But, if the profession is to rise above this disturbing tendency to accept clerkdom and is to assert itself to achieve its full impact, then ambitious, influential I.E.'s must adopt the broad perspectives suggested herein and guide industry to a higher plane of effectiveness.

PART THREE

THE POTENTIAL ROLE OF INDUSTRIAL ENGINEERS IN SHORT-TERM PLANNING

In the short term strategic planning must be translated into day-to-day operations to achieve long-term planning goals and objectives. It is not always easy to draw fine distinctions between long-term and short-term activities; there is too much interaction between them. A common view is that short-term procedures consist of little more than fighting fires, that is, single-mindedly concentrating on immediate concerns. However, the short-term aspects of planning must be performed successfully if strategic plans are to be implemented as targeted.

Success cannot be achieved if the company is constantly forced to divert its energies in reaction to immediate crises, and the only way this mode of operation can be avoided is by thoroughly planning operational systems to handle short-term needs far ahead of when they will need to do so. Short-term planning consists of designing information systems, organizational structures, analytical procedures, and methods for monitoring and reviewing all systems that will ensure their successful operations.

Short-term planning as an element of strategic planning should exist only to implement the goals of the long-term strategic plan. It would be a mistake to focus myopically on operational goals to the exclusion of corporate goals. Short-term procedures must be tested for validity according to how they enhance the company's ability to reach its broad objectives and its specific strategic goals.

Effective implementation of strategies requires data to be gathered in a timely fashion and reviewed conscientiously, and this implies that a systematic approach to doing so must exist before the right information can be collected. This is the universal need of all short-term strategic activities. It is admirable for a long-term plan to lay out a specific change in strategy as soon as conditions change enough to demand a shift in tactics; it is, however, impossible to make such short-term changes unless the requisite information systems are installed along with the conceptual plan.

This is a basic requirement of short-term planning—the need for effective information systems. Most firms claim to have a panoply of information systems, but unless they are tailored to the strategic plan, the company's operating modes, its organizational structure, its corporate culture, and the plan's monitoring needs, it will fail. One reason information systems commonly fail is that systems tend to impede their own success; the scientific statement of this tendency is called Le Chatelier's Principle. It is implicit in this observation that it is unnatural for systems to succeed without conscious effort to avoid the common causes of failure.

In the case of information systems, special pains must be taken to see that they do not simply perpetuate existing habits. Left to their own devices, corporate units will obey another well-known principle—they will follow the path of least resistance, gathering and transmitting whatever data are readily accessible. Thus, someone must design information systems based upon rational criteria derived from the strategic plan.

Whereas long-term plans can be broad and imprecise, short-term planning must account for all contingencies. The unexpected must be anticipated, and the methods for dealing with contingencies must be prepared beforehand and must include provision for coordinating diverse operational functions. The work of coordinating detailed operational procedures and integrating them into strategic plans would appear to be the province of a unique individual, someone with a broad systems viewpoint and solid operational knowledge, someone who understands broad corporate strategy—in short, a strategically aware I.E.

13
PRODUCT POSITIONING

Product positioning consists of: addressing specific markets; setting pricing policy; implementing competitive strategy; and timing product introductions, promotions, and retirements. As was discussed in Part Two of this book, product positioning requires knowledge of the types of markets being addressed, the life cycles and durabilities of products, competitors' market positions, how well production is adhering to cost targets, and accurate data for use in predicting demands. As will be a persistent theme throughout this part of the book, operationally useful information systems are the key to success in short-term planning, and this observation applies especially to product positioning. For example, consider market maturity estimation.

MARKET MATURITY ESTIMATION

At the time a market becomes officially mature, a new phase of strategic plans should be activated relative to that market. In the short term the important consideration is predicting when maturity will be reached and observing when it actually happens. Toward this end market conditions need to be constantly monitored. The company also needs some model of the market against which indicators can be compared. Ordinarily, a maturing market is modeled as tending toward a fairly static annual demand asymptotically, as suggested in the graph of Figure 13.1.

The best indicator of the market having matured is probably the slope of the demand curve; when that slope begins to flatten out, it is time to switch from young-market strategy to mature-market strategy. Timing can be very important because the change in strategy will probably imply reduction of inventories, fewer marketing resources being devoted to that market, lower production levels, and the employment of revised marketing ploys.

Assuring that demand tail-off be detected in a timely fashion should probably be entrusted to someone who understands the market, but also the information needs to be transmitted to other key management personnel. The critical need

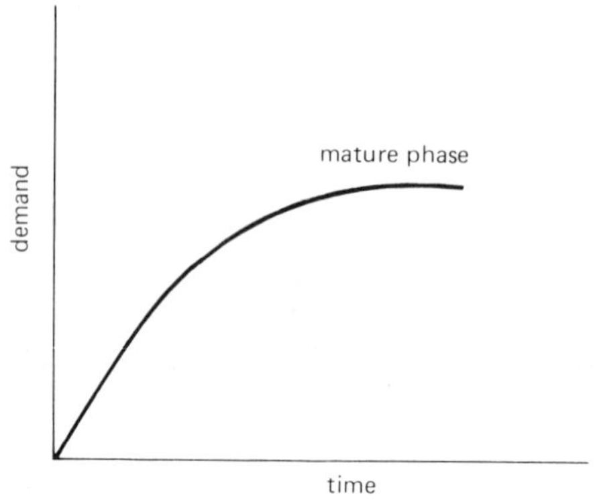

Figure 13.1. A Typical Product Maturation.

in monitoring any market changes is the timeliness of the data used for observation. Thus information must be collected from the best sources and monitored conscientiously. The structuring of the information flows is a job best done by someone with a systems viewpoint, someone much like an I.E.

Other changing conditions can dictate changes in marketing strategy and pricing. For instance, raw material prices can change unexpectedly, and, when this happens, prices must be reset rapidly or substantial losses could result. There is simply little alternative to passing along material increases, so it is important that changes of this nature be detected and included in short-term considerations promptly, which fact again implies the need for prior consideration of methods for monitoring price data.

Product strategy must respond in the short-term to competitors' tactics. If a competitor markedly cuts prices, then the response dictated by the long-term plan must be invoked in a timely fashion. This response is probably contingent on a variety of current data, such as whether the competitor's move has actually given it a significant competitive advantage. For instance, Japanese automakers have found that recent conditions have made them virtually impervious to rebate schemes employed by U.S car dealers; their market in the U.S. is simply not affected by the domestic promotions. Thus, in their case no response is required.

Imagine, however, what would happen if the Japanese dealers saw their sales drop in a manner clearly attributable to customer loss to American dealers. Then one might expect a price battle to begin. The key to making an apt response to competition is having data available soon enough to be useful. Again, the organization of data gathering and the structuring of information flows is exactly the type of work that should enlist the skills of I.E.'s.

Customer preferences may change. If they change rapidly enough a company could find itself with a large amount of dead inventory. To avoid this eventuality consumer profiles should be gathered frequently for thorough analysis. Industry-wide demands along with sales data for one's own company should be routinely gathered, studied, and reported to those who must make positioning decisions. A rapid decline of a product's sales can indicate the need to reduce production while simultaneously preparing a new product for market.

Customers may act just the opposite, showing a sudden marked increase in demand for a given product. Rapid response is also called for in this case, else sales may be lost for want of inventories. Thus, any wide fluctuation in consumer demand needs to be detected, analyzed, and reported rapidly. This requires that the framework for accomplishing this effort be established as a routine part of corporate activities. It is the kind of activity that I.E.'s are good at designing.

In the case of market maturity monitoring, all that is required is that recent sales data from the market's suppliers be obtained and studied. The key requirement is to detect demand leveling off before it is too late to liquidate excess inventories of that market's product. This collection and analysis of data must be done routinely, not just when danger seems imminent.

There are many other types of data that should be studied routinely to aid in implementation of product positioning strategies. One important area is that of production costs.

EXPERIENCE CURVE UTILIZATION

Theoretically, if a product's costs follow an experience curve's form, the constant-dollar cost per unit should decline in a fairly predictable way as cumulative production increases. Often the experience curve effect is described as being the decline of cost per unit by X percent as cumulative production doubles. This percentage may be quite large, as has been the case in the electronics industry.

If a single company attempts to employ experience curve theory in forming and implementing its product positioning strategy, it must have systems in place to both monitor cost progress and plan the next steps in cost reduction. Primary responsibility for this work should be entrusted to those whose acknowledged forte is cost reduction, they being the I.E.'s.

The strategy that employs the experience curve to position a product is similar to that of the veteran poker player who loses the first few hands by betting on such sterling cards as a pair of threes but who rakes in a large pot when his unsuspecting opponents are unaware that he is now betting in earnest. The experience curve user cuts his price to the bone, undercutting the opposition while attracting market share and name recognition for a new product; then as product costs decline down the experience curve he can further cut prices knowing that his competitors will always remain above him on the experience curve. Like the poker player he is willing to swap early losses or low profits for long-term gain.

The trick is to stay on the curve. Nothing endangers the positioning plan more than nondeclining costs. As it is with freedom, the price of declining costs is constant vigilance. Costs might decline slightly if left to do so on their own, but that is improbable because people will continue to perform tasks in exactly the same way unless they are directed to do differently. Someone, such as an I.E., must be constantly alert to cost reducing possibilities and eager to suggest and implement them.

The theory of the experience curve suggests that products can be positioned in the marketplace by planning their price decline to track cost decline [24]. Alternatively, the additional profits that result from a product with declining cost but nondeclining price can be used to develop or maintain other products. The decline of cost per unit would effectively make the product a cash cow. However, the key to any strategy of product positioning based upon the decline of costs down the experience curve is that costs must actually decline. In short, the success of a cost-based strategy rests upon the success of cost control, the activity at which I.E.'s are acknowledged experts.

PLANT-LEVEL COST CONTROL

Economic theory says that low-cost producers drive out high-cost producers. In the macrocosm of an entire competitive market, one can see that the expiration of the high-cost producers is one reason that unit costs decline. Another cause is technological change; as cheaper technologies come into use, they drive out the more expensive ones.

At the company level, it may be that a version of the macroscopic reasoning can be employed to force costs down, but it should be emphasized that total costs per unit must be used as the basis for experience curve analysis. This means that a system of fully allocated costs should be the basis for determining the total costs associated with the product in question.

Low-cost producers should also drive out the high ones at the plant level, but it may require an extensive engineering study of production to unearth changes that will produce units at lower costs. The changes in activities that will produce anticipated cost reductions are not simply predictable, but there is a core of candidates that should be examined.

Factors contributing to cost improvements may be materials, methods, production processes, or specifications. A group technology approach that reduces setup times for a number of similar products may do the trick. Reorganization of assembly operations or tighter control of standards may be needed.

The organizational structure itself may need revision. As was mentioned previously, systems installed to support a new product tend to become superfluous after a short time has passed. For example, a New Product Task Force may be assigned to help introduce a product into a new market, but this group can easily be phased out and reassigned as the product becomes accepted. This move alone will reduce the labor component of the product's cost.

Implementation of new technologies has become a frequently adopted avenue for cost reduction. NC (numerical control) machines have been a good example of technological improvements that have driven down production costs. Numerous other computer applications can and have improved operations and saved money.

As diverse as the possible means of cost reduction are, the strategic question of prime importance is not how costs are kept on target but whether they are. If cost-per-unit goals are not achieved, tactics and perhaps even strategic objectives may require revision. Thus, information systems must be established before the attempt to use experience curve-based pricing tactics to acquire cost data in usable form, and then real costs per unit must be monitored to determine whether costs are on line.

The implications of this seemingly mild requirement are sweeping. If fully allocated corporate costs are used as the cost basis for experience curve analysis, then all relevant corporate costs from the upper levels of management, through the related staff functions, all the way through production and line support functions must be reported in a timely manner. And if all these data are to be reported quickly enough to support action there must be in place an information system dedicated to this one task.

The usual corporate response to this suggestion is, ''We already have a corporate-wide, fully complete reporting system with reports issued monthly by the computer department.'' The difficulty usually accompanying these large-scale information systems is that they are so large, inspecific, and intractable that they are rarely useful for short-term applications. In addition, comprehensive systems are invariably designed to use standard accounting data, which are unsuited to providing the costs an I.E. needs to develop in such activities as tracking fully allocated costs. Monthly corporate printouts ordinarily report data that is as much as two months old, so by the time a report is almost due for release the last report may reference data about three months old, far too late for use in fast-moving short-term strategic action.

The existence of the big computerized reporting system may be a source of the corporate resistance aroused when an I.E. suggests the need for a separate cost monitoring and control system, but it is not the only source. Corporate accounting systems are set up to provide standardized data required by the Internal Revenue Service and the Financial Accounting Standards Board. The accounting systems may be additionally geared to supply some managerial accounting information, including standard cost postings and variances. However, even these systems will not provide the kind of detailed information needed to relate the resources of specific corporate functions to line activities.

The I.E. who says he needs additional data as suggested above for tracking total product costs and resource utilizations may sound as though he is suggesting the creation of a system that will duplicate existing information-gathering and reporting activities. This complaint is predictable and understandable but not true. In the operational arena timeliness and preciseness of data are essential to short-term implementation of strategic intent. The cost-monitoring I.E. must persist in developing the systems he needs if cost reduction goals are to be met.

Ordinary marketing strategies to sell established products require slightly different information. Two aspects of data gathering and analysis important to the implementation of these market strategies are those of demand estimation and determination of price elasticities.

PRICE ELASTICITIES

People generally believe that cutting product prices will inspire greater demand for products, that demand is elastic and a function of price for most products. In response to this belief, the marketing departments of companies often promote sales of products by offering them temporarily at reduced prices. In order for price-cutting to be considered an intelligent move, it is necessary to have some idea as to how much demand stimulation will result from the reduction.

Price elasticity of demand is defined by economists as the ratio of fractional demand change to the fractional price change that inspires the change in demand. For instance, a price elasticity of -1 implies that a 10 percent reduction in price will result in a 10 percent increase in the number of units sold. Say that a product is priced initially at $100 per unit, a price which yields a profit of $25 per unit. At that price assume that 400 units are expected to be sold per month, yielding a total profit of $10,000 per month. If the price is cut for a month to $90 per unit, then profit is reduced to $15 per unit. If the price elasticity in this case is -1 then the sales during the promotion month are expected to be 440 units, which would result in a total profit of $6600.

This fanciful example demonstrates the danger of arbitrarily cutting prices and the need for accurate knowledge of elasticities. It must be mentioned here, however, that elasticity is a theoretical property of future sales and only a demonstrable property for past sales, but it is important to know when the observed elasticities have become significantly smaller. If this goal is to be accomplished, then there must be an information system in place that can both acquire the data needed to reassess elasticities and to analyze and report results.

Marketing must be made aware of meaningful changes in the assumptions it makes in scheduling cost-cutting promotions.

There are several exceptions to the need for accurate data. One is in the case of a new product which has no historical database upon which to base elasticities. The other exception occurs when the profit margin is so much higher than planned price cuts that almost any increase in demand would support a price-cutting promotion. It can also be argued at times that the longer production runs scheduled in anticipation of increased sales can drive down unit cost so much that profit per unit could be assured at an even lower price.

However, in general a reasonable guess at price elasticity is needed. Wide swings in customer response to price reductions must also be monitored; they may be clues to basic changes in the market being served. In addition, significant changes in consumer response to sales tactics may be the consequence of structural changes in the economy. Whatever changes seem to show up in the customer's price elasticity of demand should be discovered and explained promptly.

A common problem at the local level lies with local management and salesmen. If sales seem to tail off more than market conditions would warrant, the fault may be in the way the product is sold. Perhaps the product is not marketed as attractively as it once was or is not as convenient to purchase. These aspects of selling might be the root cause of seemingly declining elasticities. Field inspections by marketing managers may reveal some such causes of declining sales, but the original clues to their likelihood will probably be found during the perusal of routine, sufficiently detailed sales data. No such clues will be found, however, unless data are religiously kept, monitored, and reported in usable form. The same rule should govern the cost data.

COST SYSTEMS

The requirements of cost-monitoring work vary with the kinds of products being considered, which in turn depend upon the strategic plan for the products and the product positioning tactics being employed. The tactics then dictate the role industrial engineers should play in the study and control of costs.

In a mature market the emphasis is not so much on innovation but on maintaining an already predictable cost per unit of production. In this case it is especially important that production not drift out of control, and toward this end it is important to observe and understand cost variances. If the cost base of a product, which requires a steady cost performance to remain profitable, varies significantly, this fact must be discovered and its cause found quickly.

It may be that the facilities for making the product have become worn out or that low-cost materials are no longer available. In any event the facts must be found out promptly. Cost variances can only be detected quickly if accurate cost standards are maintained in concert with routine variance reporting procedures.

In the case of a new product, costs should drop rapidly early in the product's life as procedures become routine and as startup personnel are pulled off the developmental tasks associated with startup. It is important that an I.E. be intimately connected with new products so he can learn when the product's costs begin to take on the eventual stable pattern to be expected of it, whether that pattern be experience curve decline or invariance. In either case, the I.E. needs to lay out the cost and resource standards and a system for predicting product costs and for tracking variances.

SUMMARY

Information systems in planning and control will be discussed more fully in later chapters, but product positioning strategies are strongly reliant upon cost systems discussed in this chapter. All strategically related information systems have several facets in common. They must be rapid—old information is often useless. They must be accurate. They must be to the point. The problem that all-encompassing central reporting systems present is that they are slow and do not address the specific data needs of the operational environment.

Since I.E.'s are the acknowledged cost control experts in most companies, it should naturally fall to them to structure and monitor the information systems that most strongly impact product positioning. Costs in the plant, as developed from I.E. cost predictions, must be subtracted from the corresponding sales in the field to determine actual profit margins and ascertain whether profits are in line with strategic plans.

Toward this end I.E.'s need to develop not only the cost monitoring, cost control, standards development, and variance monitoring programs ordinarily expected of them, they must also devise reporting systems that summarize recent actual and expected behavior of profits as a function of costs, prices, demands, variances, trends, and possible changes in important parameters. The primary requirement of management in the positioning of products is rapid and accurate data that can be translated into the analysis of profitability and that can motivate rapid action.

An I.E. developing procedures to support product positioning should always keep in mind the ultimate purpose of the information. A firm understanding of corporate strategic plans with respect to product positioning should help the I.E. direct his effort properly.

14
PRODUCT DEVELOPMENT

One fundamental marketing activity central to profitability is product development. Unless new products are almost constantly under development to take over when currently marketed products falter, a company can find itself left in the dust of competitors who have prepared for market vicissitudes. Short-term elements of product development are primarily coordination, monitoring, scheduling, and budgeting.

The I.E. function is not a motivator in the process of selecting the specific product developments to be marketed. This is the purview of Marketing [1] and upper management. The feedback needed from I.E. personnel concerning the choice of a product change or innovation is an analysis of whether the product is possible to make and how much it will cost to make it. An I.E. should also determine marginal profits that would be derived from new products, based upon the market demand and price predictions developed by marketing personnel.

Upper management, assured by marketing and product development personnel that a product is producible, can determine whether its production and profitability profile, as supplied by the I.E.'s, is acceptable. Then it must be decided whether the product satisfies the criteria of the strategic plan. If products are introduced on whims of Marketing or of upper management, the firm cannot be said to have a plan.

After product innovation suggestions have been judged to be economically justified and strategically sound, schedules of development activities can be prepared. Product development is similar to routine production in that the entire product development timetable explodes backwards from the date intended for product introduction, much as production schedules explode backward from promised due dates. The critical aspects of development of targeted products are scheduling and budgeting.

The timing of product introduction may be depend on the rate of decline of a product already existing on the market. Although a profile of the life cycle of the old product will have been projected, life cycle prediction is more art than science, and products have been known to hang on at acceptable profit levels long past their forecasted demise while other products have declined unexpectedly. In the former case a new product can easily be delayed, but in the latter case, it is probable that an unavoidable lapse in total sales will occur. There is, unfortunately, no way to plan for all contingencies.

However, given the best possible plan, its critical elements are the initial scheduling and budgeting followed by monitoring and coordinating of activities. These are the phases of product development that seem to need the expertise of the I.E. One might note that this short-term, product-specific work is founded on principles established at the long-term, strategic level.

SCHEDULING OF DEVELOPMENT

It is critical to product introduction that time estimates much like the standards applied to routine activities be used to schedule and cost development functions, even though these numbers may be difficult to assign. Otherwise it will be practically impossible to even guess when a product might be available to markets. This is often the problem that upsets the entire product strategy. A good example of this specific event seems to have occurred during the Christmas shopping season of 1984. Coleco had targeted that shopping period to introduce its Adam computer at a medium price for the home computer shopper. The product was distributed far too late for the Christmas impulse buying, and the product ensuingly died. To avoid just this kind of situation, I.E.'s should coordinate with the product development, marketing, and production staffs in developing meaningful development schedules.

The assumption underlying this approach to product development is much the same as that underlying scientific management—that every complicated synthesis of activities can be broken down into its constituent components, each of which can be understood and studied individually. Precedence relations can be established, and the entire process of product development can be modeled as an abstract network of activities useful in predicting and drawing up a product's development schedule.

Early in the process of theoretical product design, the design should be reviewed to judge its producibility by the production staff. Further, it should be studied for its economic feasibility by I.E.'s and verified as much as possible as to its marketability by the marketing staff. All of these efforts should be coordinated and summarized by I.E.'s for presentation to upper management. All steps need to be assigned times for evalution of total development project duration. Then the critical activities in the project should be identified by use of PERT analysis, and all activities should be modeled for this analysis, including the work of the I.E.'s themselves.

A diagram of possible planning steps in the development of a new product appears in Figure 14.1. If an estimate of the elapsed time involved in this set of steps is to be made, then estimates of the timing for each step must be acquired. The responsibility for scheduling and for making cost estimations for these steps should be given to I.E.'s, as should the responsibility for coordinating schedules between the diverse contributing corporate units.

Even upper managers must be asked to perform in a responsible and timely fashion, and this fact should be reflected in the scheduling diagram, which will then show them to be accountable if they constitute a bottleneck. Similarly, every other contributor should find himself under pressure to adhere to timetables and budgets. The fundamental accounting principle applies—budgeting is planning. Thus even the planning steps in Figure 14.1 must be included in the company's budget. One clear consequence of this stipulation is that, since a certain number of these product analyses and decisions must be anticipated in the strategic plans, the effort must also be funded explicitly as development budgets.

Budgets and schedules require databases prepared by I.E.'s, but a considerable amount of background work by I.E.'s is required to develop the time pro-

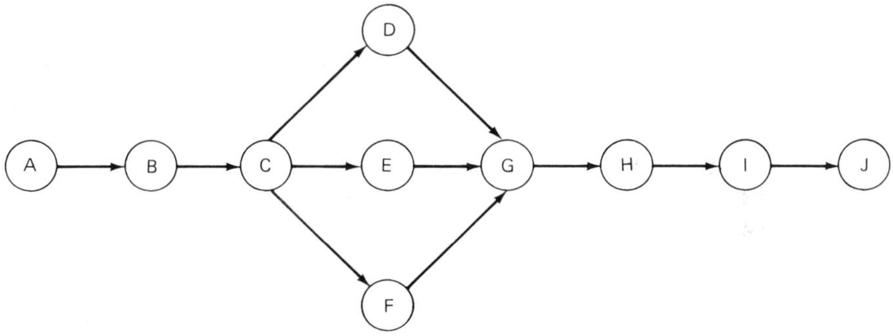

Legend:

A - preliminary market analysis
B - management strategic study
C - preliminary product design
D - in-depth market study
E - production evaluation
F - I.E. cost/profit study
G - management decision
H - product development scheduling
I - budgeting for development
J - development order release

Figure 14.1. The Preliminary Part of a Representative Product Development PERT Diagram.

jections and cost basis upon which to found the time estimates and budgets. Admittedly, these numbers may be highly uncertain, but the alternative to making such projections is to abandon the effort altogether and resign oneself to permitting unplanned, unscheduled, unbudgeted activities to determine the company's future. I.E.'s should stand firmly on the side of acquiring basic information.

INFORMATION FOR DEVELOPMENT

Figure 14.1 provides a framework upon which to base evaluation and review of the development process. The work of monitoring a PERT-like schedule consists of constantly checking whether activites are on schedule, and, then, if they are not, ascertaining how to reallocate resources to remedy any delays. The data needed as inputs and feedback for the process depicted in Figure 14.1 start with the preliminary market analysis conducted by Marketing or by an outside survey firm. This survey should specify the price range of the product under discussion and the corresponding estimate of market penetration.

These price and demand numbers can then be translated into revenues per quarter over the planning period. Both upper management in its strategic analysis and I.E.'s in their cost/profit study would need these data. Both groups would be expected to apply some version of sensitivity analysis, expecting that the obtained numbers might be somewhat optimistic. Either a mathematical variation approach or simple scenarios could be employed to test the effects of deviations of prices or demands from expectations. After management has given its initial blessing to a project, the design process, based upon product specs provided by Marketing, can begin.

Not all product development attempts will be successful. A new product or an innovation may be summarily abandoned at any stage of the process. Therefore, the strategic budget for these activities must include allowances for a percentage of development failures among the number of projects expected to be initiated.

It should also be noted that the picture of Figure 14.1 is oversimplified in that there are usually interactions between some of the activities that are pictured as distinct. For example, it is unreasonable to design a new product without input from the production sector of the company, so preliminary product design should include continuing discussion between Marketing, Engineering Design, and Production, although a thorough production study of the design will be conducted after a complete design is in hand.

After manufacturing, distribution, and sales of a new product have become routine, coordination between divisions still should be maintained. For instance, if design changes are suggested, they should be subjected to considerable scrutiny prior to being adopted. Marketing may discover that certain changes would make a product more desirable to customers. Design revisions can then be laid out by Engineering Design, and ensuingly the new design can be submitted to Production to determine its feasibility. I.E.'s can then determine the effect the changes will have on product cost, and the coordinated analysis can then be passed along for final management decision.

Similarly, cost reduction suggestions can be submitted by I.E.'s for coordinated review, and changes to improve flows can be reviewed when they are suggested by Production. These coordinated reviews are important because even seemingly innocuous changes can influence aspects of products, such as appearance, in a way that reduces attractiveness or convenience to consumers.

Information exchange between the industrial engineers involved and Marketing and Production is necessary throughout the product development effort. To effectively perform its scheduling and cost/profit studies, the I.E. needs the best updated estimates from the other functional disciplines as product development studies progress. Upper management should also interact with the separate other functions, so that strategic plans can be refined as information clarifies product prospects.

After the management decision has been made to initiate a specific product development project, short-term planning for scheduling and budgeting become the next steps. Industrial Engineering may be the best source of project managers for the product development effort. They are the people most likely to have access to and understanding of the multidisciplinary activities of new product development.

DEVELOPMENT SCHEDULING

By the time product development scheduling is ready to be done, an array of preliminary activities will have to have been already done. This fact shows that the scheduling of preliminaries is an activity that must be provided for at the the macro, or long-term, strategic level. The scheduling and budgeting of approved, specific product development activities are short-term procedures, but the ability to accomplish them is derived from long-term plans. The information systems and data must be ready for application as needed.

Prior to scheduling, time projections and cost projections for the individual developmental tasks must be already estimated. These data should have been developed during activity and capacity analyses, cited earlier in this book as the foundation of scientific, bottom-up planning.

In the case of fairly repetitive product development, such as refining the design of stereo turntables, scheduling may be a simple matter, since the new product may differ only slightly from the previous model. For instance, the solitary design change might be substituting a direct drive for a belt drive. In developing entirely new products the changes may impact schedules throughout design, testing, production, and introductory marketing stages.

The need for a variety of engineering data in development planning and schedule review demonstrates the fact that information systems to provide the data must be installed and operating at all times. This further suggests that utilitarian information systems are a strategic necessity. Since budgeting is of great importance in development planning, budget data must similarly be readily accessible.

DEVELOPMENT BUDGETING

Funds have to be taken from the total developmental budget to support any newly approved product's development. The amount of support must be based upon reasonable cost estimates per unit of resource multiplied by the resource requirements suggested by the development plan. Once the overall development schedule and its budget are established, the next important task is to monitor the actual progress of the development work to assure that it stays close to being on-time and on-budget.

The work of overseeing schedules and budgets should be performed by personnel outside the individual disciplines involved and should be performed by someone dedicated to productivity improvement and cost control. This may be the role of a project manager, or it could be assigned to someone, like an I.E., as part of general coordination duties.

Monitoring development budgets should be much like reviewing production costs and schedules. Variances must be accumulated, evaluated, and acted upon. Expediting efforts must be expended as needed to put critical late activities back on schedule. Continuous review of project steps is required to ascertain whether the work is proceeding as intended. Costs are much more difficult to control than schedules because of the tradeoffs between the two. Expediting of late work costs money, but there is no way to retrieve spent funds after they are gone.

One of the most bitter complaints of project managers is the lack of adequate procedures to determine whether the budgets of one-of-a-kind projects are on schedule. The kind of tool that might fill this need might be a PERT-like financial model of projects that will permit them to be readily reviewed for current vs. expected levels of expenditure. The accepted PERT procedures are of some value in reviewing time schedules, but both time and money schedules are difficult to review with great accuracy, and for similar reasons.

The time schedule is hard to review with accuracy because it is usually difficult to tell how much longer it will be until a partially completed activity is finished. In reviewing the expenditure schedule it is often difficult to tell how much has been spent on the job and how much is yet to be spent. In addition, estimates of both expense and time requirements of a one-of-a-kind project are uncertain, a fact that compounds the problem.

The surest method of project control is tight coordination of project activities by project management personnel intimately familiar with the pertinent details. It is unlikely that automatic algorithms will ever suffice to hold unique projects on track; people will be required. They must be objective people, capable of exchanging information between disciplines.

COORDINATION AND MONITORING

Coordination of activities is as important an element of product introduction as any other activity. Given a strategically dictated date for product introduction, every other schedule explodes backward from that management-selected date. This date also determines the beginning of the advertising campaign and the start of the marketing push. Distribution points are notified of the resulting schedule of distribution and shipment sizes to expect. Product development staffs are informed when they must be finished with product debugging, and production is informed as to when it is to produce a specified number of units. Next, the required units are posted on the master schedule, which in turn results in the scheduling of production runs and the ordering of materials.

All of these activities must be monitored and checked to assure that coordination has in truth been achieved. If long-range planning is effective, the information exchange and coordination system should have long been installed as part of business as usual, but no system will run itself, which is why monitoring of activities is necessary. In addition, there should be interlocutors to talk between disciplines to iron out any problems that arise in the heat of the moment.

Until a new product has established itself as a routinely scheduled item, it will constitute a special case; no habitual procedures relating to its production, movement, and sales will be established until considerable experience has been acquired. Until it does become a routine item, coordination and monitoring of its progress will constitute an addition to existing procedures. These additional duties should be absorbed into established information systems as rapidly as possible, which will reduce the new product's resource requirements and allow efforts to be redirected, perhaps toward additional product development.

A coordinator cannot be chosen from one of the disciplines that regularly compete with one another for corporate attention. It is therefore logical for someone who views his domain of concern as that of overall corporate profitability to oversee the early developmental activities. An ideal individual for the job is an I.E. An excellent organizational approach would be to make an I.E.

responsible for developmental scheduling, budgeting, and coordination and monitoring. Whoever directs the work must deal with the steps likely to be involved in launching a new product.

Product-launching dates project backwards from the day chosen for exposing the new product to final consumers. This date fixes the time at which product must reach distribution points. The schedule must be exploded backward from distribution point arrivals. Figure 14.2 schematically displays the activity sequence. In this picture the activity labeled Preliminary Product Development includes all 10 steps cited as constituting the diagram of Figure 14.1. Given the date chosen for product introduction at the distribution points, the subtraction of delivery lead times should produce production order due dates.

These order due dates are then entered into the production master schedule. Production run scheduling, in turn, explodes backward until an initiation date for production order release is reached. The order release date must be preceded by the completion of the development and testing phases for the product. Unless final specs for production are available at the time of the order release, the entire production and and distribution schedule will slip.

The preliminary development schedule has already been discussed, and it must be included in the initial plan to consider the new product, both in scheduling and budgeting. Early in its progress, initial product commitment is ten-

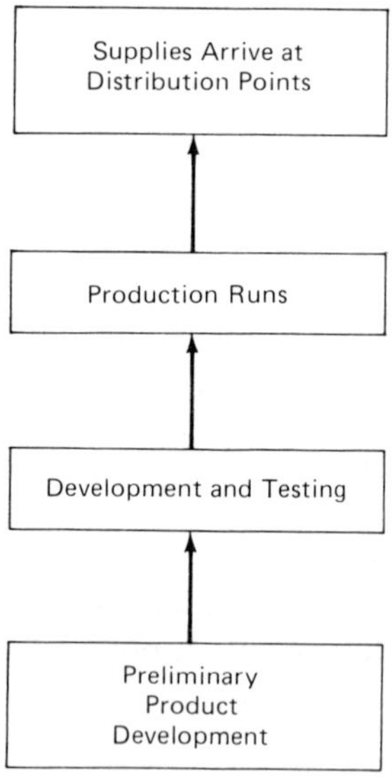

Figure 14.2. Macroscopic Schematic of Product Development and Distribution.

tative, and it is only when management has decided to proceed that definite scheduling and budgeting can be performed. Thus preliminary product development steps must be budgeted as part of the total, continuing development activity.

After the schedule has been laid out for a new product all the way through market introduction, costs of the various steps can be guessed, probably from data describing similar past activities. Then the next chore is the monitoring of developmental and production activities to assure that they remain on schedule and under budget. Information exchanges between departments necessary to keeping the project on track must be coordinated throughout the schedule. Again, it seems reasonable for an I.E. to be in charge of these activities.

Early production data must also be scrutinized carefully so that accurate standards can be devised. Production of the new product will eventually go more smoothly than in the early stages, but it is important to begin tracking the product's manufacturing profile early. It is also important to convert the startup information systems into data components in the routine production information systems as soon as possible. The development of reliable resouce requirements and the standardization of methods must be included among the I.E.'s duties along with the other aspects of shepherding of new products.

SUMMARY

Product development crosses disciplines and therefore can benefit from objective assistance of the kind I.E.'s can provide. Too often product development does not have sufficient control exerted upon it and becomes a randomly directed corporate activity. Lack of upper management involvement, lack of prior planning and budgeting, lack of strategic foresight, and lack of the organizational structure needed to manage development are the usual causes of a weak product development function. I.E. assistance in structuring development activities can help avoid some of these common weaknesses.

Industrial engineers clearly should not have major responsibility for market innovation ideas; this is a Marketing activity. Neither should I.E.'s be strongly involved in development work itself; this is the work of specialists, also. But the scheduling and cost aspects of the development process can benefit from I.E. expertise. Also, since product development is a multidisciplinary function, it requires specific attention to the coordination of the disciplines. As an objective party, only interested in seeing the firm succeed, an I.E. seems to be an ideal choice for someone to direct that coordination.

15
PRODUCT MIX

At the strategic planning level of decision making, product mix of the short-term variety is given only cursory attention. In macroscopic considerations, a statement of broad product mixes over fairly long planning periods facilitates the test of whether production capacity can nominally satisfy the plan. At that level product mix analysis is not of primary importance but only one aspect of the broader activity analysis suggested earlier in this book.

However, at the operational level of activities and in the short term, product mix takes on considerable importance. A particular plant, trying to plan its production schedule over a period of a few weeks will tend to view strategic criteria as mere guidelines. For instance, policy may dictate that the plant make enough of each of ten major products to satisfy major customers without incurring stockouts more than five percent of the time. Within these broad guidelines, there is a good bit of flexibility.

Although short-term product mix strategy is admittedly far removed from the years-long foresight of corporate strategy, the ability to plan production efficiently is a large determinant of corporate productivity and profitability. The scientific choice of product mix at a production facility should be guided by the peculiar factors of primary importance to operations at that facility.

PRODUCT MIX CRITERIA

There are several possible product mix criteria, each of which may be reasonable in different contexts. The two most common criteria are cost minimization and profit maximization. Other possibilities for optimal goals are maximum throughput, minimum materials handling, and maximum capacity utilization. All of these objectives require approximately the same information to model.

As has been mentioned, product mix analysis is inherently short-term and thus will ordinarily deal with only variable costs as opposed to the fully allocated costs suggested for the full analysis of corporate activities. An obvious reason for ignoring fixed costs is that the long-term plans have already considered and determined the fixed components of costs over the short periods considered in product mix analysis.

In addition to variable costs, a product mix analyst will need labor, machine, and material standards for each potential product along with capacities of each production process. The exact sequence of steps per product will be needed, too. Over the periods to be planned, the expected product demands and the constraints imposed by strategic criteria must also be available.

The truly long-term aspect of product mix analysis is that it should be performed by use of an analytical model that was installed as part of implementing the strategic plan. There are a number of possible product mix analytical tools, but the most successful is linear programming (LP).

An LP model seeks to maximize a criterion (such as profit) function within the flexibility left to the plant by imposition of strategic and operational constraints. The underlying rationale motivating the employment of LP models is that operating systems create unpredictable interactions between subsystems that can severely restrict profitability. If these interactions remain poorly understood, this fact can be responsible for incorrect corporate pricing and costing procedures, which in turn can reduce profitability. The cure for this problem is obtained by providing management with the appropriate bottom-up analysis, using such tools as LP. Let us examine a simple product mix example to see how poor information can restrict profitability.

Consider three fictitious products that can be made from the same materials at the same production facility. Call the products A, B, and C. A study of forecasted demands and inventory policies has been interpreted into de facto upper and lower production limits for the three products. The pertinent information is listed below:

Table 15.1. Product Demands and Standards.

Product	Weekly Minimum Demand	Weekly Maximum Demand	Labor hrs/unit	Material lbs/unit
A	50 units	100 units	3.0	42
B	100 units	150 units	4.0	45
C	150 units	200 units	2.5	40

Sales Prices per Unit: A—$250
B—$300
C—$280

Resource Costs: labor —$12.00/hour
material—$4.00/lb.

A small amount of calculation will reveal that, assuming all manufacturing units can eventually be sold, the profit margins per unit (after variable costs) are:

A—$46.00
B—$72.00
C—$90.00

The highest profit per unit goes to product C. The clear implication is that C with a P/V of 32 percent would be the best product to make in bulk. The implied product mix would consist of 50 units of A, 100 units of B, and as much of C as possible.

However, a more careful investigation was called for when it was discovered that not even the minimum demands could be produced. It was then found that the apparent bottleneck was machine time. Careful I.E. work showed the machine to be nominally available for about 100 hours per week and further showed that machine time requirements per product are as follows:

A—requires 0.5 hours of machine time, per unit.
B—takes about 0.3 hours of machine time per unit.
C—needs a full hour of machine time per unit.

Machine time constrained the number of units that could be produced and changed the profitability picture from what had been assumed. In terms of scarce machine time profitability is:

A—$132 per machine hour
B—$240 per machine hour
C—$90 per machine hour

Thus, product B actually has the highest profit per unit of scarce resource, indicating that it should be produced with priority.

Economic logic shows that, given machine time as the scarcest resource, the most profitable products per machine hour are, in sequence, B, A, and the C. If minimum product demands are dropped to zero and maximum limits are retained, the product mix is:

150 units of B
100 units of A
  5 units of C

for a total profit of $15,850, obtained by multiplying the units by their prices and deducting the costs of material and labor.

The lesson of the above short example is that one must be sure of the true values of resources involved in production before using them as bases for production and profitability decisions. Improper evaluation of profit margins leads a company to have only a fuzzy idea of what its product mix should be and why. The costing of scarce resources is a fundamental economic principle and derives from the fact that only scarce resources have economic value. Thus unless a constraint truly constitutes the plant's dominant resource constraint it should not be used to establish product mix.

The question of how profitable a specific product is becomes important primarily at the strategic level of decision making. There managers need I.E.'s to provide them with accurate rational bases for individual product decisions, but, in the mechanical process of product mix modeling, estimates of single-product profitability are not needed. For the small product mix problem presented above, the LP formulation of the problem would be:

maximize $(250A + 300B + 280C − 4MATERIAL − 12 LABOR)
subject to:

$$3.0A + 4.0B + 2.5C = \text{LABOR}$$
$$42A + 45B + 40C = \text{MATERIAL}$$
$$0.5A + 0.3B + 1.0C \leq 100.0$$
$$\text{LABOR} \leq 1000.0$$

All variables non-negative.

Demands would also need to be included as constraints in the model, but it was impossible to produce all of the minimum demands, so the demand constraints would probably be removed and then added on experimentally until the best profits were achieved. The demand constraints that provide the solution cited above would be:

$$A \leq 100$$
$$B \leq 150$$
$$C \leq 200$$

In the LP model, no estimates of profit margins are required because the only costs that enter are variable costs, and they must all be used in an LP model only as direct costs. This procedure assures that when requirements of materials are totaled for the selected manufacturing schedule, the model does not use some imputed value of resources to products but instead uses their known prices. The LP model will produce an optimal product mix, based upon resource utilization, capacity constraints, costs, and revenues without resorting to profit/unit estimation; but any such estimation process, if it is to be of value, should give a total profit for a given product mix that closely imitates that of the LP solution. The LP mix is:

100 units of A
150 units of B
and 5 units of C

which uses all 100 machine hours but only 912.5 labor hours and 11,150 pounds of material for a profit of $15,850.

There are other modeling approaches to product mix analysis, most of which essentially employ simulation models. The advantage LP holds over simulation is that the simulation model only studies a specific product mix at a time while the LP model considers all possibilities at once. However, no matter what modeling procedure one uses to study short-term product mix, the approach used should also provide routine plant loading information.

PLANT LOADING

The basic assumption an accountant makes about production costs entered into ledger accounts is that they are based upon fairly accurate standards and that, when the actual expenditures are recorded, their variances from the predictions of the standards will ordinarily be fairly small. That is, the accountant intends to use standards as reasonable predictions of real costs. Similarly, a plant manager hopes to be able to predict just how a planned product mix will affect his operations, and he hopes to obtain a good prediction based upon those same standards.

Regardless of whether a plant uses an optimization model to plan its product mix, it still needs a production planning model that predicts the consequences of any given mix. At some plants there is never any question of which product mix to make—the order simply comes down from on high. However, once the product list has been given to the plant, a loading model is needed to predict the plantwide consequences of the given mix and to help prepare the plant's master schedule over the current planning period.

There are many practical loading models available, from LP to simulation to the general spreadsheet software that can be obtained for use on many of the currently popular mini- and microcomputers. The model needs to be able to take the product demands over the planning period and interpret them into cumulative resource requirements, preferably broken down by master schedule time buckets, which are usually either weeks or months.

Given the resource requirements, the master schedule can be massaged and rearranged, and it can be used to develop specific schedules of resource acquisition, labor force scheduling, and production scheduling that will satisfy order due dates. If there is no flexibility in the due dates, which would occur in only a few instances, then the loading model would be of little use, since the master schedule would be predetermined.

One point as to terminology needs clarification. The expression ''master schedule'' means different things to different people. To upper management and Marketing the master schedule consists of the set of quantities ordered by customers and the dates they are promised. To production personnel the master schedule is more likely to be viewed primarily as the production schedule, including not only customer due dates but also production order release dates. As an example consider the master schedule worksheet shown below.

Table 15.2. Master Schedule Worksheet.

Time Bucket	1	2	3	4	5	6
Order Due Date			30		20	20
Planned Order Receipts			20			
Inventory 20@	20	20	10	10	10	10
Requirements					20	20
*Production Order Release		20	20			

* product lead time = 3 time buckets
@ initial inventory

This is the worksheet for a single product that carries with it a requirement of a minimum inventory of 10 units. Starting inventory is 20 units, and prior production order releases are expected to produce 20 units of finished goods three weeks into the planning period. The 30 units promised customers in the third period appear to be an old order, but the requested quantities in periods 5 and 6 are new.

Working through the time buckets reveals that to satisfy due dates and sustain the required inventory level, there will be twenty units of production needed in each of periods 5 and 6. Since production lead time is three time buckets, the production order releases must be backtracked three weeks from the requirements for end products, so the need in period 5 translates into an order initiation in period 2, etc.

Upper management may consider that the top line in the worksheet represents the master schedule; this is the ''bottom line'' mentality appropriate to such personnel. The plant manager will consider the entire set of quantities of interest and may view the worksheet as the master schedule after it has been amended to include the inventory consequences of the production schedule that has been derived. Here is where the plant loading model may come in handy.

The implicit scheduling criterion used in working through the production times is that production is to be scheduled as late as possible without violating inventory standards and without delaying due dates. This is not necessarily the operative criterion a plant manager would use. He may know that delaying certain orders can be done with impunity. He may believe that level plant loading is of more significance than just-in-time production. Or, he may be intent upon maximizing throughput, even if this means building inventory with off-peak capacity.

For the purpose of examining the plant loading concept let us attach resource requirement standards and capacity constraints to the single product in the master schedule. For simplicity assume that the product spends one week in each of three departments—1, 2, and 3, in sequence. Descriptive numbers are:

Resource Standards

Department	1	2	3
Labor hours per unit	1.2	2.0	1.7
Materials (lb) per unit	8.5	5.2	1.1
Machine hours per unit	4.1	3.2	5.3

Resource Capacities

Department	1	2	3
Labor hour limit per time bucket	80	80	80
Material (lb) limit per time bucket	200	100	50
Machine hour limit per time bucket	100	70	160

For the planned order receipt lot now in its second week of production and the two newly initiated orders, the resource utilizations during the planning period for the stated master schedule are as follows.

Plant Loading Analysis

Time bucket	1	2	3	4	5	6
Labor, Dept. 1 (hrs)	—	24	24	—	—	—
Labor, Dept. 2 (hrs)	40	—	40	40	—	—
Labor, Dept. 3 (hrs)	—	34	—	34	34	—
Materials, Dept. 1 (lbs)	—	170	170	—	—	—
Materials, Dept. 2 (lbs)	104	—	104	104	—	—
Materials, Dept. 3 (lbs)	—	22	—	22	22	—
Machine, Dept. 1 (hrs)	—	82	82	—	—	—
Machine, Dept. 2 (hrs)	64	—	64	64	—	—
Machine, Dept. 3 (hrs)	—	106	—	106	106	—

The plant manager can glance at this output of data and see that the existing master schedule is comfortable and only overextends the materials availability for department 2 by four pounds in each of periods 1, 3 and 4, a problem he can easily expedite. However, if he is not happy with this set of resource utilizations, he can revise the master schedule and examine the new output from the plant loading model.

The example just completed is only a one-product problem. A standing plant loading model would include all major products and all major resource requirements. Pareto's Law again dictates that not every product and not every resource must be included in the model.

The product mix aspects of the plant loading model require resolving capacity conflicts. As soon as two or more products are placed in the master schedule, they compete for resource availability. Juggling the timing of the production schedule is a way to resolve resource conflicts, but it requires both cognizance of due date criteria, inventory criteria, capacity limits, and operational criteria. There are computerized procedures that can help resolve many scheduling difficulties. Management needs some kind of standing plant loading model, but humans should make the final scheduling decision.

One interpretation of a model even as small as the the one-product model presented above is that it shows which resources have slack capacity that can be diverted to other uses. For instance, during hard times some plant managers have been known to use slack to solicit extra business, say in a sideline such as metal fabrication. Or extra capacity might be diverted to product development or methods improvement research. Knowledge beforehand of capacity requirements can be put to many valuable uses, including building inventory.

INVENTORY IMPLICATIONS

The choice of a short-term production mix affects all facets of production, including inventories. Ordinarily, a heavily produced item will require similarly large quantities of work in process and raw materials to support its high output level. In addition, production to some degree is parts-oriented, including subassemblies, parts, and fabricated product components, which should be produced in quantities sufficient to take advantage of the cost economies of long production runs. Since parts tend to be produced in large lots, parts inventories will naturally accumulate in response to the product requirements.

Another peculiar effect of certain product mixes is that all inventory levels may tend to ratchet. Manufacturers tend to schedule production runs according to fixed lot sizes, so inventories have a tendency to either be very large or very (vanishingly) small. This effect may persist merely because lot size determinations have never been questioned. A good I.E. should question any production policy that creates peculiar conditions in the stockroom. Perhaps flexible lot sizes should be invoked, or, even better, lot-for-lot production could be adopted, thus minimizing work in process.

Another unusual inventory-related feature of some product mixes is the behavior of intermediate stocks. As has been mentioned, intermediate stocks are those materials, parts, or subassemblies that can be used in several ways—in several finished products, for example. It has been found that such stocks tend to follow a seemingly random inventory level. Until the solution to wildly varying intermediate stock levels has been obtained, plant managers are well warned to at least be prepared for the effect to occur.

Inventory considerations may play a large role in determining the production mix. For instance, corporate policy may strongly prohibit stockout conditions among the major products. In such a case the determination of the product mix may descend from inventory considerations. The dominant criterion might well be stockout avoidance, which policy would impose large safety stocks.

For an I.E. the problem that such an inventory rule presents is the disruption of smooth production scheduling that the rule causes. One approach to minimizing the disruption is to pick a level of stockouts that is minimally permissible—say allow no stockouts 95 percent of the time. If management agrees to such an approach, the I.E. can build a simulation model to simulate past demands over time while choosing production levels that achieve the accepted inventory criterion. In the case of a new product, marketing's sales projections will have to take the place of past data in simulating product orders. By running the model and revising the production levels iteratively, a production mix can be selected that legislates sufficiently against stockout events. Study of the assigned production levels can then allow the I.E. to derive safety stock policies for the important products.

EXTENSIONS OF PLANT LOADING MODELS

Once a plant loading model is in hand, there are two rewarding applications of the model and its conceptual basis. These are the kinds of extensions of problem-solving tools that I.E.'s are trained to look for. The first application is to the selection of a multiple-plant product mix.

A single-plant loading model can be extended to a two-plant model by combining the two models and adding linking constraints. A good example of a linking constraint is the upper limit on total demand for a product made at both sites. An optimization approach to distributing the product's manufacture might mix production at both sites, or it might assign 100 percent of that product to one site. The only reason the model might not choose the longest possible production run at a single site would be that the single plant does not have enough capacity to complete the total demand.

A two-plant model will produce a number of unexpected revelations about capacity utilization and product mix. The reason for this fact is that the more components a production system has, the more unexpected interactions between components occur.

Multiple-plant models are large but are put together in the same way as are two-plant models. Multiple-plant production models are similar to multiple-site distribution system models. They are built in much the same way. The distribution model answers questions peculiar to sending units of product to diverse sites.

The difficulty faced in assigning deliveries of products to distribution sites is determining the mix of quantities. Often a company will have a reasonable prediction of overall period demands for a product, but regional demands are difficult to sort out within the confines of the macroscopic projection. Models to decide how many units to ship where are like multiple-site production models in many ways and can be built and analyzed in a similar fashion.

Basic information of the sort familiar to I.E.'s needs to be obtained as a first step in model building—such data as warehouse capacities, past inventory performance vs. sales, delivery lead times and costs. Distribution options such as regionalizing distribution need to be defined. In a simple model, distributions to distribution points can be simulated by choosing arbitrary fractional assignments of the total demand to the various distribution points. The quantities delivered can then be computationally compared with predictions of site sales (developed from past records) to determine which sites would tend to be oversupplied or undersupplied by the given distribution mix.

Thus, a distribution policy can be devised that would depend upon the product and forecasted overall demand. The policy can be revised over time—fine-tuned—and the computational model can be revised as better data is acquired.

An LP model that seeks to minimize the costs associated with distribution can also be constructed. Costs might include transportation costs, holding costs, and even a fictitious cost attached to running out of stock. A variety of distribution models for the analysis of multiple-site distribution can be assembled, but they are all similar to the models specifically constructed to study product mix. And product mix models are just a special case of activity analysis models.

SUMMARY

Many problems are isomorphic to product mix problems, which generically seek to choose a mix of activities within the known constraints that gives either an optimal value or a very good value of a criterion function. The data required to build any such model are essentially the kinds with which I.E.'s are fully familiar. In fact, the basic data for these models would probably be either gathered or coordinated by I.E.'s.

No matter whether the model's criterion function is cost or profit or the number of stockouts or the size of inventory, the underlying purpose is the same. That purpose is to use information about the operations in question in a way that will produce intelligent, objective examinations of operating policies. It may require a good bit of intelligence to translate computational results into operational policy, but without objective, bottom-up analysis policies would be largely just guesswork.

The strategic significance of short-term models such as those directed toward analyzing product mix problems lies in their very existence. Part of the corporate strategic plan should be concerned with laying out information systems, analytical models, operational systems, coordinating systems, and monitoring and evaluation systems. All of these systems and models should be tied together into comprehensive analytical tools.

The end result of the installed systems should be that the methodology will be ready for dealing with all kinds of short-term implementation problems. A good example of the application of the described systems methodology is in the area of choosing and governing methods of production.

16
METHODS OF PRODUCTION

The broadest decisions governing the methods of production are made at the strategic level. The long-term commitment of capital for facilities and equipment define a company's range of production options for many years. Yet within these limits are a number of short-term options for the utilization, maintenance, and replacement of facilities and equipment that directly affect costs and productivity.

Some of the process decisions can be made with a short planning horizon, especially if they require only small investments, but most of the short-term management of methods of production must be prepared for long ahead of time during strategic planning activities. As usual, the existence of the necessary information systems is the sine qua non of effective action.

Among the most important information structures is that contained in the production planning and control systems to be discussed in Chapter 17. While the use of these systems requires some specific elements of production design, design generally lies outside the production control function. However, the design aspects of production planning and control must be worked out before operational systems can be employed, so these aspects are discussed here.

DESIGN FACTORS IN PRODUCTION SYSTEMS

Many production control systems are now available from vendors, and they all promise to be panaceas for whatever ails profits. The catch to these systems is that they all require some degree of automation of functions. Whether manual control of a machine is to be replaced by numerical control or whether a clerical function is to be computerized, it should first be determined that the automation is a good idea and, if so, how it will be accomplished.

Consider the seemingly simple automation of collecting the machine times used to fabricate a lot of material. A device is hooked up to each machine and records exactly when the machine runs. The only remaining difficulties are the translation of the data into elapsed time and the association of that time with the correct lot of material. It often happens that this procedure does not work well. Either the on-line sensing of the machine does not perform accurately or the machine operator forgets to log the starting times for the lot.

Whenever such a seemingly simple automation function fails, the cause is usually inadequate preparation. The common error is installing the new automation procedures too abruptly, without checking them against the manual procedures they replace. If there were no workable manual procedures in the first place, then the automation would be doomed from the start. For automated systems to succeed they must replace existing, but less efficient, systems. Success depends on preparing for system installation properly.

Successful system automation should require the efforts of a cooperative team of specialists. An excellent choice as a team coordinator would be an I.E., someone who understands the goals of the automation, who has a feel for operational realities, who understands cost flows, and who desires productivity improvement.

A team might have need for a computer programmer, a production supervisor, workers familiar with the jobs to be automated, and any other relevant specialists, such as accountants or manufacturing engineers. The team would develop the hardware and software configuration of the desired automation process and plan for its implementation. An I.E. would assign cost figures to startup and routine operations and would draw up installation and operational schedules.

If management were to accept the automation plan, costs, and schedules, the team's plan would probably be implemented. The implementation plan developed by the team would then go into effect. A considerable portion of the startup time for new automation procedures should be spent on training the personnel who must interact with the system in order for it to work [26]. To plan the education of the workers requires that the planning team fully understood the information flows and human actions needed in the system.

A common system in which this necessary groundwork is often missing is that which should ensure quality control. The usual error committed by American companies is inspecting only final products. This procedure shows that management does not really understand the production process nor does it know where quality control problems originate. The principle violated is that quality cannot be inspected into a product [15].

Quality comes from managing each step properly—from selection of suppliers to acceptance sampling to training manufacturing workers in the correct procedures in their jobs to monitoring critical production steps to testing and experimenting throughout the manufacturing sequence. Technical knowledge of correct feeds, speeds, maintenance schedules, and quality indicators is the basis for training a quality-conscious workforce. But training the worker must be preceded by management's acquiring knowledge of the worker's job.

Similarly, any automation of systems requires the same kind of thorough before-the-fact knowledge of the system being designed. However, even a manual system that is to be newly installed will require careful research and training first. Consider, for example, the Japanese kanban or just-in-time approach to routine manufacturing.

The premise of the kanban system is that it will run the same way all the time. The first requirement that must be met before a kanban plant can be operated successfully is that there be a stable product demand. The simplest way this criterion can be satisfied is for the kanban plant to aim at supplying only a fixed part of the company's total output of its product.

Given fixed demand, a secondary premise that a Japanese kanban plant must assume is that there are to be no in-plant inventories, not even buffer stocks between production line stations. This requirement differs from the usual U.S. approach of allowing buffer stocks almost everywhere, even in plants that are viewed as just-in-time operations.

Given the two basic assumptions, the Japanese kanban plant is laid out and scheduled so as to meet its specific demand goal. No attempt is made to laboriously balance line loadings. The fastest operation in a line is likely to be slowed until all steps operate smoothly. Smoothness, reliability, and avoidance of inventories are the goals of the kanban operation.

The reliability of every individual and every machine are painfully clear in this system. No problem can afford to go untended because it closes down the whole plant. The Japanese involved in such operations feel that the scrutiny each step receives pays off in increased reliability, accuracy, and quality.

What is very clear is that management must know exactly what it is doing to be able to implement a kanban operation. Contrast this type of plant with the usual American plant burdened with order backlogs, excessive scrap and rework, and scurrying expediters, and one can see the problems a lack of preparation can produce.

While design of information, planning, and control systems is of high importance to effective governance of production, another aspect of manufacturing that has not been given enough attention is that of plant availability. An increase of only a few percent in a plant's availability can mean a considerable increase in its output and the attendant revenues it garners. The design considerations that are important to this topic are those of redundancy and maintenance [12].

REDUNDANCY AND MAINTENANCE

The design of information and production systems will primarily determine how routine operations are conducted, but it will have little to do with controlling downtime costs. Downtime to a large degree is a function of system design, and once a plant is in operation its design may be difficult to revise. However, increasing uptime of critical manufacturing steps can reduce the costs incurred by downtime, and there are two direct ways to increase uptime. These are the use of redundancy and preventive maintenance.

Redundancy's effect on plant availability is easily demonstrated. Consider one machine that must be operational if the plant is to function. Assume that the machine has a 0.75 probability of being operational at any time, or, stated another way, the machine will be down about 25 percent of the time. If another identical machine were in place, able to rapidly take over the primary machine's duties during the primary's downtimes, then the combination's function would be unavailable only 6.25 percent of the time.

The question that must be answered to justify component redundancy is whether the benefits of greater machine availability outweigh the capital costs of installing the additional equipment. If the addition increases cash flow, then the marginal cash flow can be evaluated to determine whether the rate of return on the marginal justifies the incremental investment. Alternatively, redundant components might be justified simply because they smooth production flows.

Smaller pieces of equipment ordinarily have redundant components, although they may not be viewed as such. Extra supplies of light bulbs, bearings, gaskets, conveyor belts, rolls for rolling mills, and wiring are commonly stocked to provide for rapid replacement of failed or worn items. The fact that these units are hidden away in the storeroom makes it less obvious that they constitute redundant design. The obvious redundancies in designs ordinarily appear as parallel flows, as demonstrated in Figure 16.1.

The primary component, denoted as A, is ordinarily in operation, but when it goes down the backup unit A' takes over as primary. When A has been placed back on line, it serves as backup until A' fails. In computational algorithms that simulate system availability the active component is labeled as A, and the other is labeled A' without regard to which was named A originally. If, as earlier supposed, both components have a probability 0.75 of being available

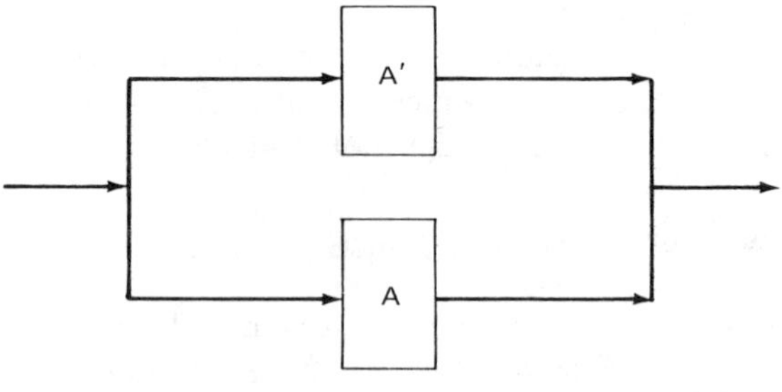

Figure 16.1. Parallel Installation of Redundant Components.

when called upon, the following logical derivation shows how the system availability increases to 0.9375.

P(A) = 0.75, the probability that the primary unit is up when it is needed
P(A') = 0.75, the probability that the backup component is up when it is needed

$$
\begin{aligned}
\text{P(either A or A' or both are available)} &= P(A) + P(A') - P(A)P(A') \\
&= 3/4 + 3/4 - (3/4)(3/4) \\
&= 24/16 - 9/16 \\
&= 15/16 = 0.9375
\end{aligned}
$$

If a replacement unit is able to be rapidly brought from the storeroom and used to replace the failed component, then this situation is very little different from the ideal state shown in Figure 16.1, which assumes physical adjacency and automatic switching upon failure. As was mentioned, smaller units ordinarily have replacements stored for them, but large, expensive units often do not.

At most production facilities little thought has been given to redundant design. Usually management acquires designs for a plant, designed without parallel components (thus minimizing total first costs), builds the plant, stocks replacements for small items, and simply suffers long delays when large components fail. The stage at which redundancy should have been thoroughly thought out was at the design stage, although it may be necessary to consider it later as well.

To determine whether a parallel component is desirable as a backup for a machine or work station or major machine component, it must first be learned whether its failure will severely affect plant availability. If an expensive item does not slow up work upon its failure, it can be left without a backup. An I.E., studying the advisability of redundant design, would not want to violate Pareto's Law by concentrating on components that cause little trouble .

Once the I.E. has drawn up a short list of potential redundancy locations, he should ascertain the total costs and benefits of installing backups. The most common benefit is increased revenues due to higher output during the increased uptime. The probable costs are those associated with purchasing, maintaining, holding, and operating the redundant components. If the incremental cash flow over the anticipated life of the backup produces an attractive rate of return on the investment in redundancy, the backup should be installed.

There are cases where design constraints prohibit the installation of availability-enhancing backups. For instance, some facilities are built with no extra floor space for additional equipment. Another common limitation on redundancy is a shortage of storage space for those smaller units that must be inventoried but do not have to be physically adjacent to the units they back up. An investigating I.E. would find redundant designs inapplicable in such instances.

Even when redundant designs are not possible or practical, there is another approach to increasing availability that an alert I.E. should consider. This approach is the scheduling of preventive maintenance (PM). The employment of PM has much the same positive effect on availability as does parallel design and can be viewed as a reasonable alternative to use of redundant components in some cases. Better still, PM can be implemented at any stage of operations—not just during the planning period prior to plant construction.

PREVENTIVE MAINTENANCE

An effective PM program relies strongly on accurate maintenance data. This is a stringent requirement, given the way most firms approach maintenance. A quick look into the maintenance records of most companies is disconcerting. There is next to no data there!

The usual purpose of keeping maintenance records is to charge off maintenance wages. The records are intended to show roughly what the maintenance staff has been up to. As a result, the maintenance records are none too detailed. Commonly a maintenance slip will record the date along with the following:

Equipment: Pump
Status: Broke
Action Taken: Fixed it

This is really no information at all. An I.E. trying to design redundancies or trying to develop an intelligent PM program simply must have decent maintenance data. This point is also important to the topic of replacement, discussed below.

The logic of a PM program is that, if a unit can be repaired systematically prior to an expected breakdown, downtime can be drastically reduced and productivity improved. However, for this approach to work it must be known which components are the important ones to schedule for PM, what repairs or replacements are to be anticipated, and when PM should be performed to prevent unexpected failures. All of these pieces of information must come from reliable past data.

The I.E. who is trying to bring rationality to maintenance scheduling will probably find his first task to be installing an adequate reporting system. At minimum he needs to know which components have failed and when, what caused the failures, what was required in the way of parts and labor to effect repair or replacement, and how long the maintenance work took from start to finish.

Introducing such a detailed reporting sytem will assuredly be difficult. Maintenance personnel homogeneously dislike paperwork and generally suspect people like I.E.'s of trying to catch them goofing off. Only if the I.E. expends the effort to educate maintenance personnel to his purpose is the reporting system likely to produce accurate data.

To demonstrate how maintenance data can be used to plan a PM program consider the very simple case of a motor that freezes up and must have its bearings replaced often. The information the I.E. planning the PM program needs are the mean time to failure (MTTF) and mean time to repair (MTTR). These numbers indicate how often to expect a failure and how much downtime to expect a failure to incur.

The I.E. puzzling out this problem will probably work from two data sources—plant logs and maintenance records. The fact that logged downtimes are found to correlate with motor failures would establish the motor as a maintenance item worth considering for PM. Accurately kept maintenance records should allow the required average times to be calculated. Exactly how the preventive bearing replacements are to be scheduled depends on a number of factors.

Some facilities schedule PM during an off shift. For instance, if a plant routinely operates two shifts a day, the third shift can be reserved for planned maintenance. Another approach to PM is to simply schedule downtime for PM, the reason being that several different PM procedures can be peformed at once, thus virtually ensuring the avoidance of several sources of unexpected downtime.

The exact time to schedule PM for a given item, such as the motor mentioned above, would similarly depend upon diverse factors. Its remedial work might be scheduled prior to its expected time of failure or it might be serviced at the same time other nearby items are due for attention.

The long-term characteristic of both designing redundancy and planning a PM program is that they require accurate data concerning system and component availabilities, failures, and repairs. The availability of the data depends upon the existence of information systems and procedures that routinely acquire, summarize, and monitor the data. The installation of these systems and procedures must be at least implicit in the strategic plans or else the systems may never exist. Instead, such folksy mottos as, "If it ain't broke, don't fix it," will describe the way maintenance is handled.

An amazing fact is that reducing downtime may be the key to profitability, yet it is rare that an organized approach to its reduction is attempted. It does require organized procedures. Inventories must be adequately stocked to supply backups and replacement parts as needed; this requires supervisory attention and money. Schedules must be drawn up and data must be maintained. These activities require coordination and review. Again, these organizational facets of availability-enhancement programs should be described directly in the corporate strategic plan, else they will not even be budgeted.

A closely related operational topic is that of the replacement or retirement of capital stock. This topic has considerable strategic significance.

RETIREMENT AND REPLACEMENT

There are two basic and quite different ways of viewing capital investments in plant and equipment. One is that they must make money to justify themselves. The other view is that, once the money for them has been spent, it is gone—a sunk cost. The two views lead to differing views of strategy and different approaches to replacement and retirement.

If an investment has been justified on the basis of expected cash flows over a protracted period, management may find it very difficult to pull out of the market the investment addresses. On the other hand, if management treats past expenditures as a dead issue, it is much easier to avoid sending good money after bad.

In fact, the first viewpoint is in error. Past commitments of capital should not affect strategic planning of future commitments except to the extent that they are educational. Only future profitability is at issue. Discontent with management's tendency to cling to outdated investments may be partly responsible for the recent approach to strategic planning spawned by consulting firms that label some old investments as "cash cows."

A cash cow, according to the consultant's conventional wisdom, must not receive significant new investment; it must be milked dry and then allowed to die. That is certainly not a policy marred by excessive sentiment, but it is also a strategy that at times has been abused. A cash cow is nominally a segment of a business that is not growing and that is serving a market that is not growing. A cash cow has no hope of increasing market share. It is typical of a mature and possibly saturated market. Yet pulling out all significant capital investments to that segment of the company almost surely dooms it to extinction, whether the move is warranted or not.

Before a firm's segment is labeled a cash cow and allowed to atrophy, a hardheaded analysis of its ability to make meaningful returns on future investments should be performed. There are many companies that would like nothing better than to have a stable, reliable division that could be counted on when all else fails.

The standard consultant advice is that profits from the cash cows that are being phased out should be diverted to exciting new products that show considerable potential for growth or to recently developed products that have already shown themselves to be highly successful. The drawback to this advice is that it tends to emphasize the speculative portion of the company's investment portfolio while diminishing its low-risk support base. This advise is of little risk to the consultant but of considerable risk to the advisee.

The lesson herein is that a reasonable mix of speculation and stodginess is good business. Therefore, each segment of the business must be analyzed as to its worthiness for investment quite apart from considerations of the past. Some companies are, after all, nothing but cash cows. A company that prints bank checks is, due to the growth of electronic banking, in a declining market, but if it does good work it may well survive while other less efficient printers fail.

The unsentimental view should always be taken toward such capital investments, including those that relate to equipment retirement and replacement. The common procedure for timing new equipment that is taught to I.E.'s in college engineering economy courses requires that an item be replaced at that time when its total of annual operating and capital costs exceeds estimates of that same total for the new item that is to replace it. Comparisons are usually made by projecting costs for each year up to some planning horizon, discounting them back to present, and then annualizing them. This approach ignores past expenditures, which is good, but it requires that several critical assumptions be made: (l) that the old item and its replacement do the same job; (2) that they do it the same way; and (3) that costs are important to the decision, not profits.

These assumptions might be tenable if a machine is to be replaced by an identical machine. Even so, the third assumption above might be invalid. A preferable method of calculation of replacement/retirement alterntives should be performed by using the cash flows associated solely with the equipment under consideration. This method consists of calculating each year's projected cash flows directly associable with the equipment in question during the planning period. The cash flow (after tax profit plus depreciation) should be calculated for each alternative: replacement during the current year; during the next year; during the year after that, etc., for each replacement alternative. Alternatives might consist of replacement with the same kind of equipment, replacement with similar equipment, replacement with a new technological advance.

The use of cash flow instead of profits in the calculation is the preferred approach because this method stresses the importance of funds freed by depreciation, which are much higher for new purchases. Again, the capacity for performing such calculations must be provided for as part of long-term information planning. Only when marginal revenues and costs associated with individual pieces of equipment can be isolated can marginal cash flows be estimated. An I.E. involved in replacement analysis should also have a hand in developing the needed cost and revenue data.

The fundamental basis for cost estimates is resource standards. Revenue estimates have to be based upon similar numbers. For example, if three machines are used to manufacture a product, sales revenues from the product can be allocated to the machines according to machine hours per unit of production.

Alternatively, one could prorate all revenues equally to each machine as a basis for revenue standards. The second method will probably not yield very different replacement decisions than the method based on machine hours unless the relative commitment of machines to the product vary considerably.

SUMMARY

The coordination of operational procedures is steadily becoming more dependent on computational procedures. The very choice of such procedures is of long-range significance because they will decide how things are to be done and how procedures are changed. All systematic approaches to planning and control, design redundancies, maintenance, and retirement require up-to-date information that is routinely gathered and reviewed. These are the kinds of systems that should be instigated, designed, and implemented by I.E.'s. This is suggested because I.E.'s readily understand the production and economic tradeoffs that underlie choices of methods of production.

All of the problems discussed in this chapter are similar to those of the prior chapters in that their solutions rely upon operating and information systems that should be installed at or shortly after the strategic planning stage of activities. The following several chapters deal in more detail with specific operational systems and their adaptation to automated versions, which appear to be becoming more and more necessary to high-efficiency manufacturers.

17
INTEGRATED PLANNING AND CONTROL SYSTEMS

Economists have recently been studying what appears to be a new trend among manufacturing businesses—a decline of inventories. An explanation of this apparent trend is that managers have been made wary by business declines in the near past that caught them with excessive inventories. Another different, but related, explanation is that companies have steadily increased their use of computerized planning and control systems that allow them to work more efficiently, thus requiring lower safety stocks. If this explanation is the true one, inventories have become smaller as the result of a structural change in the way manufacturers operate.

The motivation to reduce inventories provided by periods of high carrying costs is just one stimulus that prods businesses to refine their operating procedures. Another is technological advances, which have steadily decreased the importance of direct labor in manufacturing. This permanent change in the way manufacturing works has caused management to take a different view of how operations should be governed. The new management philosophy stresses automation, computerization, and integration of activities. The forces that drive modern manufacturing can be demonstrated by the increasingly adopted integrated control systems that are directed toward achieving the strategic goal of just-in-time manufacturing.

JUST-IN-TIME

Just-in-time production has been discussed at times in the business literature as though it had been invented by the Japanese. In fact, the Japanese in large part have simply implemented American management principles and the advice of such Americans as Deming [9] and Juran [15]. As long ago as the turn of the century Frederick Taylor [36] and other Americans were drawing up scientific management principles that aimed at achieving just-in-time production. These principles are still valid today but largely ignored.

Nearly every job shop in America intends to devise schedules that satisfy order due dates as promised. Success in this endeavor requires simple just-in-time scheduling. American management deviates from this production philosophy, especially during periods of rapid business expansion and low interest rates. The tendency is to build inventories, a procedure that was more readily justified when work was labor intensive.

Especially in large-scale plants, and particularly in smooth-flow operations, high setup costs led to the pursuit of economies of scale, employing large facilities and long production runs. America's long dominance of international markets also caused American management to take a laissez faire approach to quality control and technological innovation in the larger industries such as autos and steel.

These contributing factors diverted business practice from its just-in-time philosophy, but times have changed again. International competition is fierce, and America's product quality has been challenged. Few industries are now labor-intensive, and setup costs due to newer automation and computation tools are being dramatically reduced. The rapidity of technological advance has made investment in large facilities seem ill-advised. At the same time inventory costs have risen until they threaten to dominate a company's periodic expenses.

Into this environment comes greater opportunities for returning to just-in-time scheduling in the form of integrated planning and control sytems. The philosophy that implicitly guides the operation of these systems is important to understand, especially for I.E.'s, since it must conform to the corporate strategy and culture of any company that opts to use them. Planning and control systems are much like facilities in the sense that, once they are purchased and installed, they dictate how the company is to operate for years.

These considerations should be weighed at the highest strategic decision level. Locking a company into the fixed operating philosophy dictated by integrated planning and control is not a decision to be made lightly, nor should it be delegated. Only upper management, in the context of overall strategy and accountability, should make the decision. As has been suggested throughout this book, it is nevertheless imperative that the decision be based upon the soundest technical analyses obtainable. It is at this point that the strategically aware I.E. should enter the picture.

The I.E. must ascertain how the just-in-time philosophy of the usual planning and control systems can adapt to the strategies and operational realities of his company. The new planning and control (P&C) systems will be intended to replace existing procedures. Whether the P&C tools will work is largely dependent on whether the procedures to be replaced are working. If not, the replacement will probably not work either. If existing procedures work, then the planning and control systems have a chance to work.

Ordinarily, modern planning and control systems require considerable use of computers. Although the term "information system" seems to have fallen from vogue, to a large extent that is what a planning and control system is. It essentially processes information that describes current conditions, massages that data according to algorithms that depict how facilities are supposed to operate, and plans out activities as tightly scheduled as possible.

A large amount of information, such as product flows, standards, and capacities must go into running the planning function. The control function relies on information, also, but of the current variety—possibly even real-time data. The company installing such a system must be willing to generate and maintain the required data. Moreover, it must be willing to accept the discipline that just-in-time planning and control demand.

Discipline is an absolute prerequisite to making the system work. Say that a production planning and control system is to be installed. One of its main purposes is to allow close tracking of production orders. If the code for a job is punched up on a system terminal, the job's current status should spring to the screen immediately. If the system is to be of any value, that status information must be very close to correct, which in turn means that the data and assumptions used to build the planning and control system's database cannot be permitted to be arbitrarily revised on whims of personnel in the front office or on the plant floor. This requires discipline. And this is perhaps the trait at which the Japanese excel [22].

Before discipline can be installed in an operation, that operation must be understood fully enough that it can be managed well without elaborate aids. Operations that are not fully understood will not work when equipped with expensive planning and control systems. There are even activities that are clearly outside the realm of disciplined control.

Consider pure research. A scientist working to develop new special-purpose ceramic materials can scarcely be scheduled into specific activities by the hour. His main job is to think—a most undisciplined activity. Similarly, artisans performing high-quality woodwork cannot be constrained to hard deadlines.

In manufacturing operations, however, there are not many activities that would not benefit from a dose of discipline. Even so, only those that are clearly adaptable to P&C discipline should be put under the governance of a P&C system. Let us now consider what aspects of a manufacturing operation an I.E. should view as candidates for computerized P&C.

In broad terms, a completely integrated planning and control system is best viewed as an array of interconnecting computational modules. One possible arrangement of P&C modules is shown in Figure 17.1, which includes the information flows and monitoring functions of accounting and sales departments.

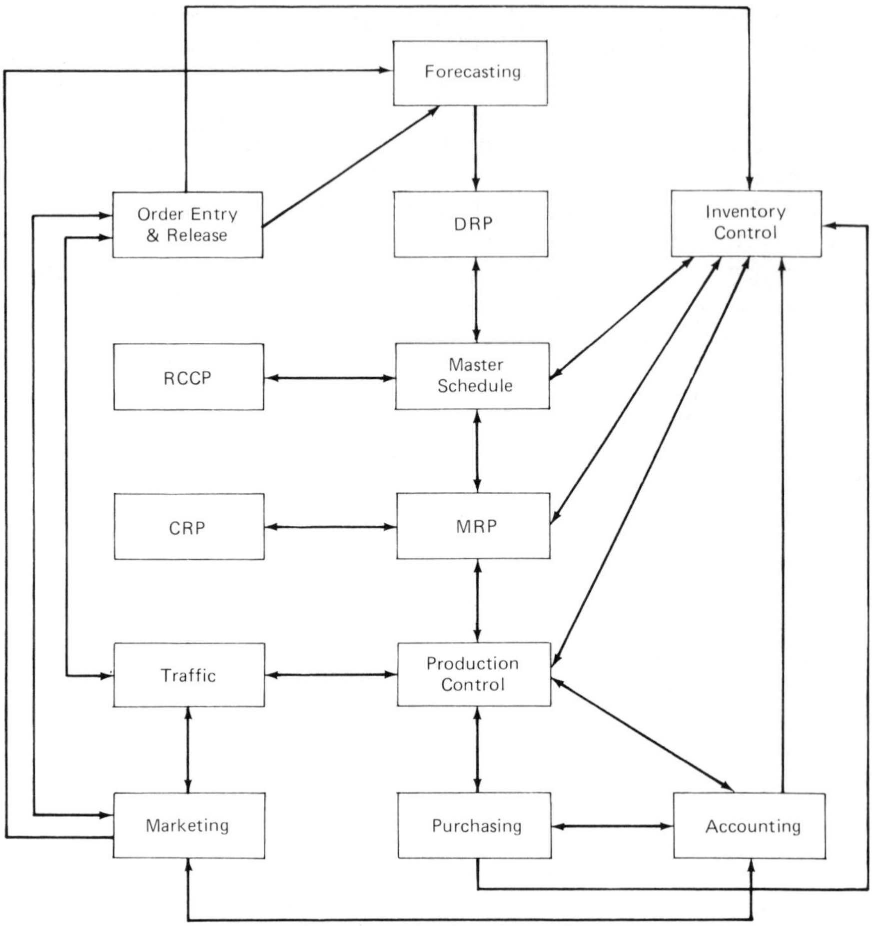

Figure 17.1. Integrated Closed-Loop Production Planning and Control System.

PLANNING AND CONTROL MODULES

The array of interconnecting modules shown in Figure 17.1 is far too extensive to fit the needs and capacities of most manufacturing firms. Some subset of the pictured P&C components might be applicable, however. A small firm that primarily does specialty work might have need for no more than inventory monitoring. Larger firms might or might not care to include Accounting or Marketing as part of the loop. On the other hand a totally integrated system would need to include such additional functions as Finance.

A short description of the possible role of each module, be it stand-alone or interconnected, will help clarify the module functions. The first segment of the system determines the levels of demand that are to drive production. Depending on how these levels are set, one or both of the Order Entry and Forecast modules will be involved.

Some companies use a demand forecast to plan their production runs and then work to sell what they produce. This approach usually builds abundant stocks of finished goods inventory, but the production schedule itself functions as though the forecast consists of firm orders which must be filled just in time.

Some companies work to satisfy firm orders only. This is common at a job shop, which may build no finished goods inventory at all. Other firms use some mix of forecast and firm orders to set demand levels. In any case the algorithmic modules naturally operate in a just-in-time manner, developing schedules from demand levels and due dates.

The Distribution Requirements Planning (DRP) module may interact strongly with the Forecasting and Order Entry modules, revising macroscopic demands into regional demands or doing just the opposite— aggregating regional forecasts into a single demand level per product. The DRP segment may even stand alone at the front of the P&C system as the demand generator. This module must explode regional delivery schedules backward across transit and distribution lead times to produce due dates for the production runs.

Due dates for products to reach final consumers represent the master schedule to marketing and upper management, but to production personnel the master schedule is more accurately described as the production order release dates because these determine the work schedule. The Master Schedule module of a P&C system may or may not be divorced from the Inventory Control and Rough-Cut Capacity Planning modules, but one sure function of the Master Schedule algorithm should be to check the preliminary master schedule to assure that the scheduled quantity due dates can nominally be met. If not, this algorithm should revise the schedule until it can be met.

Although it may be included as a part of the Master Schedule module, Inventory Control is a function that not only tracks inventory levels but reschedules production orders when inventory criteria are violated. Rough-Cut Capacity Planning (RCCP) may also be included as part of the Master Schedule function, but as a separate algorithm it calculates approximate capacity utilizations of manufacturing resources and reschedules any production orders that exceed capacity.

After a revised master schedule has been generated from demands tempered by inventory considerations and capacity limits, the schedule is ready to be exploded backwards through a Material Requirements Planning (MRP) module to develop specific production requirements. This was the original P&C computational module upon which the broader ideas were built. It has become fashionable to substitute different words while keeping the MRP acronym. Some people refer to the entire closed-loop P&C system as Manufacturing Resources Planning. Some refer to this entire system as MRP II to distinguish it from the original MRP.

All this convoluted use of language rightfully implies that there is something fundamental about MRP that is merely expanded by adding modules. The principles of MRP are the principles of the entire closed-loop system: just-in-time production based on accurate information.

MRP must be based upon accurate bills of materials if it is to work at all. Many firms have found that their operations yield confusing results from MRP algorithms and have abandoned their intentions to install one after abortive attempts. The common causes of failure are instructive.

In some cases the bills of materials start out wrong and stay wrong. MRP tries to explode a product back into its sequential fabrication, acquisition, and assembly steps, attach lead times to each step, and cut orders for acquisition of outside supplies. The MRP module must interact with inventory information to determine how much material must be ordered. A wrong bill of materials results in either the wrong things being ordered or the wrong quantities.

Sometimes the company's Engineering Department changes product specs and the word never filters down to the MRP's stored bill of materials. Sometimes lead times are far off and are never corrected, or they may become less accurate due to unreliable suppliers. Production steps may have fictitious standards assigned to them, further distorting lead times.

It should be noted that there may be conflicts between criteria at various levels of the integrated P&C system. It may not be possible to resolve all such conflicts within the computational system. In this case a decision support system (DSS) can interface with the system or the capacity for interjecting heuristic priority rules into the system may be required. In general, it is improbable that the successful installation of even a single P&C module can be accomplished without tailoring the system and training personnel.

These kinds of problems, well known in MRP systems, are less well-known in the larger closed-loop systems, but it is certain that the same kinds of problems can be expected in them. For this reason any I.E. considering whether his firm should install any or all of the described modules would do well to find out more about the pros and cons of previously installed similar systems.

Following the MRP module in Figure 17.1 are the Inventory Monitoring, Purchasing, and Production Control modules. The MRP module initiates purchases, scheduling of setups, scheduling of fabrication, scheduling of assembly, scheduling of inventory withdrawals, and calculations of specific capacity utilizations at the detailed level. Possibly as an internal aspect of the MRP module and possibly outside of it there will need to be a Capacity Requirements Planning (CRP) module.

The CRP performs the same function as the RCCP but at a more detailed level. The resource utilizations per time bucket are calculated from the schedules produced by the MRP and compared with available capacities. Necessary changes in due dates must be communicated to Order Entry/Release so that customers and distribution points can be informed.

Detailed inventory control must also be performed, either within the MRP/CRP segment of the system or in the Inventory Control module. This is a very important planning and monitoring function. However, the MRP module is often set up so that it orders materials in a highly automated fashion without much human review. Thus, review must be built into the system or specifically assigned to a human.

Production Control is nominally the most important module of all. It oversees schedules and inventories. It checks whether due dates are apt to be met, and, if they are not, it initiates expediting procedures. It is difficult to envision this module being fully automated, but automation can help a production manager acquire the information he needs to keep operations on track.

It is not unreasonable to have Accounting and Marketing tied into the information network of the system. It is certainly valuable for Marketing to know whether a prized customer is being supplied on schedule. If Accounting is prepared for automated job costing, then its efficiency can be enhanced by being included in the information flow. However, the more nonmanufacturing modules in the system, the greater the information demands and the greater the likelihood of breakdowns or error.

In outline, orders are submitted over the planning horizon to production; production is scheduled and completed; Traffic is notified and ships out products according to DRP instructions; everyone is notified that orders have been filled; and costs are tallied. There is a considerable amount of detail omitted throughout the above discussion, some of which will be supplied in ensuing chapters. Some general points about integrated systems in general, however, should be made here.

INFORMATION FLOWS

A computerized system is an information system. It is not the physical system being coordinated but a mathematical representation of that real system. Some representations may be fairly accurate, such as reported real-time flows of materials; and others are reasonable approximations, such as time standards and forecasts of demands. Some elements of the model are arguable, such as costs assigned to various indirect activities.

Each module requires an algorithm that imitates the human logic the module replaces. It also must be equipped with required established criteria, such as priority rules for choosing between conflicting alternatives. Each module also requires a database of such numbers as standards and lead times, and each must operate upon input information—operational data or outputs from other modules.

It is clear from this characterization of the P&C modules that this type of computerized system is an information system. Therefore, no such system should be installed in an environment that can not provide the full information the system requires; either existing procedures should be maintained or a computerized system should be tailored to the application.

Even when a P&C module nominally seems to suit an operating environment, information must be steadily monitored and updated. Critical data, such as costs, lead times, lot sizes, product specs, standards, and capacities, must be reviewed frequently in order to keep the computerized model current.

It must be noted that "constant data review and updating" does not appear as a module in Figure 17.1. Even though this is an extremely important part of using P&C systems, it must be viewed as external—a part of system maintenance just as physical maintenance is external to the production system. This observation suggests that a careful critique of what a P&C system can and can-not do is necessary.

MANAGEMENT AND COMPUTER SYSTEMS

A dangerous but prevalent claim made by managers at every level is "We have a system for that." This claim is too often made when the system is merely a computerized system, a mathematical model that deals with symbolic representations of real systems, not with the real systems. The key point is that the real systems, be they people, machines, or algorithms, must be understood as they really are and then managed realistically.

Computerized systems require the same kind of supervisory attention as do real components of a manufacturing firm. The basic concepts have already been presented in discussions of activities in the prior chapters. These system components need coordination and monitoring. The systems themselves need to be reviewed to determine how they are working and what resource flows are involved. And all supervisory activities must be anticipated and planned for prior to the installation of any additional systems.

I.E.'s seem to be the natural individuals to prepare for use of P&C systems because they have a cost-conscious attitude and a full understanding of the operating environment that the systems are intended to improve. However, a common error made by management is to entrust the selection and tailoring of a P&C system to personnel from the corporate computer department.

It is a major management error to allow computer professionals to gain control over production systems. Computer personnel are only obliquely interested in how production systems work and do not view P&C systems as production systems—they look on them as computer systems and feel that the primary criterion for these systems is that they mesh well with all the computer systems under the aegis of the computer department.

A strategically aware I.E. with a systems viewpoint understands that the computational aspects of the P&C system should mesh well with the interacting computer systems, but he also knows that this is not the dominant criterion for P&C systems. The real criterion is that they facilitate an improved ability to perform just-in-time activities throughout the profit-making and cost-incurring chain of corporate activities.

SUMMARY

Computerized P&C systems appear to be the wave of the future, their prospects being continually improved by the advent of more powerful hardware and software coupled with declining computer costs. However, the practicality of these systems must be verified prior to their installation. Merely because it seems fashionable to install them is not a justification for doing so.

A good I.E. can coordinate the efforts needed to assure that workable, cost-effective systems are implemented. In some industries, such as continuous flow facilities, entire plants are essentially computer-operated. Even job shops that primarily feature operations on a variety of machines have become steadily more computerized through the application of group technology and numerical control. But in most manufacturing environments minimal automation has been achieved, which suggests that most computerized systems now in use are primarily information systems rather than manufacturing systems. It may be that management is not yet fully educated to the uses of modern computers and software.

However, I.E.'s study these topics extensively during their formal educations, so I.E.'s are well prepared to lead the way in educating management to the virtues of computerized systems as P&C tools in manufacturing and in implementing those tools. One such P&C system, discussed in the next chapter, is one aimed at improved distribution and reacting to changing needs in the short term.

18
DISTRIBUTION PROCEDURES

The dominant strategic aspects of distribution were summarized in Chapter 10. They are inventory policy, distribution policy, transportation policy, cost policy, and service policy. These policies define the parameters within which short-term activities must operate.

It must be decided at the strategic level how short-term decisions are to be accomplished and what the tools for doing so will be. Policy decisions must define information structures, responsibilities, goals, and operating criteria. An implementation plan should also be developed at the strategic level.

This is not to suggest that upper managers actually devote their time to detailed planning instead of broad policy development. However, every part of the strategic plan that calls for things to be done, especially things that must be done routinely, must be prepared for in detail, else they will not be done. Personnel assignments, schedules, budgets, and continuing responsibilities must be specified.

No matter what kinds of systems are used to distribute products, they must be adequately prepared for and then adequately maintained. Let us consider some specific aspects of distribution systems that were suggested in broad terms in earlier chapters and demonstrate the part that an alert I.E. might be expected to play.

DISTRIBUTION MIX

If strategic policy is sufficiently developed, the general philosophy underlying providing goods to customers will not be an issue in short-term distribution planning. Distribution points will also be known quantities. The short-term questions to be faced are what lot sizes to ship and which transportation modes to use.

Ideally, corporate policy would be set in such a way that these decisions appear easy to make, but even the best long-term policies will not obviate the need for short-term reaction to unforeseen conditions. Dealing with these contingencies is the role of short-term strategic planning.

Say that corporate policy has set a specific number of units to be shipped to each distribution point the week of 2/12. That week comes and available units are 29 percent short. This is where short-term reaction is needed. There must be a readily available mechanism for determining whether all sites receive reduced shipments or whether one site receives no shipment or whether some other approach is warranted.

This example suggests that a manufacturing plant should have a performing distribution model that can deal with exceptional cases. I.E.'s understand how to go about constructing such models. In this instance a model that is part art, part science would probably serve best.

It is reasonable to suppose that Marketing is interested in the mix of products sent out and has good ideas as to what the mix should be. If out-of-stock conditions are expected to be frequent, a coordinating I.E. might suggest that marketing employ either a decision support system (DSS) model or a goal programming model to help avoid it.

Both of these kinds of models require intuitive insights from experienced individuals familiar with the situation the model addresses. A DSS is essentially a simulation program stored on a computer that can be adjusted to present the results of possible stockout remedies suggested by the knowledgeable user. A goal programming model might contain weights assigned to stockouts at each distribution point so that the weight given to shortages at a site indicates its relative importance. These weights can be varied by the program user to investigate shifts of stocks from low-priority locations to high-priority locations in the model solution.

Some problems arise at distribution sites because distribution is a complicated activity, involving numerous products and destinations. On the other hand, shipping is made complicated by unavoidable physical limitations; manufacturers shipping directly by rail often find themselves with distribution problems because not all customers can be reached via direct routes. A reasonable assumption is that most companies have at least intermittent distribution difficulties.

SIMULATION MODELS

Let us now examine the suggested models. First, the simulation model. This model requires data similar to that Marketing uses to make its demand projections. In particular it requires predictions of product demands over time at each distribution point.

The model would be expected to produce estimates of the consequences of any particular distribution mix, so its output should consist of data such as the following:

site demand units supplied inventory shortage

for each site and each time period of interest. The main points of interest to be read from the output would be how well the chosen distribution mix avoids overstocking while also avoiding shortages.

If corporate policy has consolidated several distribution sites for the purpose of reducing its shipping complexity, the individual site data are still needed to properly simulate product shortages and their consequences. Given projections of the site-by-site consequences of a given distribution mixes of products, some selection criterion can be applied to adopt a specific distribution mix. A good criterion might be ‘‘minimize total anticipated missed sales.‘

The steps in the algorithm are:

1. Generate expected sales of the products in question at each distribution point for each time bucket in the planning period.
2. Select a distribution mix for shipping from available stocks and project shipments during the planning period.
3. Deduct anticipated sales from the sum of inventories plus planned shipment receipts at each site.
4. Convert any negatives (shortages) calculated in step 3 into absolute values and total them over all sites and all time buckets during the planning period. This total is the criterion quantity, ‘‘number of missed sales.’’
5. The individual running the program should now either accept the current shipping mix as ‘‘good enough’’ or revise the mix and run the model again. Iterations continue until a mix is selected.

This type of procedure is said to be a DSS because the decision and even the alternatives are provided by a knowledgeable person, not by the algorithm. A DSS is a good approach to problem solving when the situation to be addressed is too complex to be modeled completely. For instance, the program operator may be aware of highly changeable conditions that are not easily programmed into the simulation but which must be considered in selecting a distribution mix.

A DSS might even be the basis for assigning stocks to distribution points routinely. For example, given macroscopic product demand forecasts over the planning period and the knowledge that the total demand could be produced within production capacities, a planner could use the DSS model described above to develop a delivery mix that would avoid or nearly avoid projected local stockouts.

The data needs of a simulation model used as a DSS are not extensive, but careful planning is needed to ensure that they will be available. The most difficult numbers to acquire with accuracy are the sales projections. They will probably be based both on past sales and on assumptions about how site-by-site sales are distributed.

Sales may have to be represented by an estimated probability distribution. The simulation model can then be run using sales generated randomly from the distribution. Running the model numerous times can then give average sales data that provide an idea of expected stockouts. Ensuingly, the sales averages can form the basis for the evaluation of the distribution system. If Marketing routinely develops estimates of expected site-by-site sales, they can be the demand data needed to run the DSS model.

Two other prominent data needs in distribution planning are regular counts of inventory levels and verifications of product deliveries. These topics will be discussed in some detail shortly. At present it suffices to indicate how they fit into a simulation model. A good guess as to currently available inventory and the number of units of products anticipated to arrive at each site are needed to determine whether expected sales can be accommodated. Verification of deliveries at the sites must be reported to alllow central updating of field inventory estimates.

The key fact that forces a manufacturer to prepare for difficult distribution choices is that conflicts are sure to occur. The diverse objectives that constitute the basis for industrial strategy sometimes call for actions that cannot be taken due to resource limitations. As a result, choices must be made. Modeling procedures specifically designed for dealing with multiple criteria may be invoked to adjust tactics in such cases.

GOAL PROGRAMMING MODELS

A simple method of making choices is to assign permanent priority rankings to activities according to their importance to the corporate strategy. However, permanent priorities are not ordinarily practical. Most manufacturers are forced into ad hoc decisions because relative priorities are dependent upon the exact conflict being faced and cannot be predicted before the exact situation arises.

Some corporate planning systems explicitly allow for revision of priorities and appoint someone knowledgeable of strategic objectives to regularly adjust the priorities used in making decisions. This is just a slightly more mechanized version of intuitive decisions made by small businessmen in their everyday conduct of business. The difference is that most manufacturing operations are too complex to allow one person to make all the decisions required. In addition, the time required to make such decisions "by hand" is often prohibitive. These considerations have led to the use of computerized planning and control systems that incorporate the consideration of priorities into computation and planning.

For instance, an inventory control program ordinarily is equipped with safety stock levels that products are not supposed to fall below. However, if a production capacity constraint makes it impossible to restock products below their limits, the question arises as to which product should be restocked first. To prepare for just such a problem, some inventory control programs are programmed to make choices based upon priority rules. A manufacturer probably assigns his best product the top ranking in the program, and, whenever a choice of which product to restock has to be made, the computer program automatically chooses the highest-ranked product.

This conflict resolution feature of some P&C algorithms is useful in the special situations to which they apply, but it is also handy to have a general-purpose conflict resolution tool around. One tool that can be applied to conflict resolution is mathematical programming, and a specific variety of this tool is goal programming.

Goal programming attempts to resolve conflicts mathematically according to priorities supplied by the program user. In trying to reallocate short supplies the priorities can be assigned to distribution points to rank them relative to one other. The mathematical model for the distribution problem can be designed as follows. The criterion is:

$$\text{Minimize} \quad w_i\, S_i + w_2\, S_2 + \ldots + w_k\, S_k$$

where w_i is a weight assigned by the program user to a shortage at site i
S_i is the shortage in delivery to site i

For each of the k sites a demand equation would be included in the model's constraints:

$$I_i + D_i + S_i = \text{predicted demand}$$

where I_i = inventory on hand
D_i = product quantity to be distributed.

Input information for the planning period would include the anticipated inventories on hand at the sites and the quantities of products to be shipped to sites. The relevant constraints would be included in the model along with user-selected weights.

The model attempts to find the distribution of shipments, given the number of units available, that will come as close as possible to satisfying each site's demand. The specific distribution mix chosen will be that which minimizes the shortfall at the locations assigned the high w weights. The higher the priority of a distribution point, the higher its assigned weight.

It is not always easy to make direct interpretations of the weights used. Often a goal programming user will assign seemingly meaningful weights only to discover that the resulting distribution mix is not acceptable for some reason. The weights must then be revised until a usable solution is found.

In practice a goal programming model performs much like the previously described DSS. The user iteratively supplies inputs and studies outputs until he is satisfied. For the DSS the user supplies the mix to be evaluated and mentally weighs the simulation results. The goal programming model requires inputs of specific weights, but it then produces an optimal mix for those weights.

The two procedures are quite similar in outline and attempt to imitate what a trained individual would do to solve small problems manually. The reason for using a model is that manual calculations tend to be time-consuming, and well-trained people are better employed making judgments from computational data than performing mundane arithmetic.

Any system for resolving distribution conflicts requires an information system to provide the basic data. As one can see in the partial formulation of the goal programming model, two important pieces of information it must have are inventory levels and delivery confirmation.

MONITORING INVENTORIES AND DELIVERIES

How many items are in inventory? When were items shipped? When were items delivered? These are fundamental questions important to almost any kind of organization. A related question is, "Where is everything now?"

Many companies have gone to the current high-tech approach for keeping track of order lots bound for shipment—bar coding. In this procedure every order is stamped with a bar code. Upon order completion its bar code is read into a computer, which in turn notifies the Order Release module of integrated P&C system that the order is finished. Order Release then sends a message to Traffic to schedule shipment.

The bar code is again read when the order is loaded, at which point it is labeled within the monitoring system as "in transit." Upon arrival at a distribution point the code is again read into a computer and transmitted to the monitoring algorithm, which in turn labels the order "delivered." If orders are being shipped to final customers, the shipper can only wait for the customer's notification of delivery, whatever form that takes.

The bar code procedure is interesting in that it not only shows what can be done with computers, it also delineates the information flows necessary to an efficient distribution system. In effect the computer procedure tries to answer the essential questions about production, inventories, and deliveries. No matter what system a manufacturer uses, the information necessary to run things right will be the same, not only when orders were received but how many items are in stock.

Inventory counts are essential to smooth operation. Only if fairly accurate finished goods inventories are known can requirements be forecasted for the distribution points, so both central and distributed inventories must be monitored closely. The older the stock counts are, the more uncertainty attaches to the production/distribution plan.

It may not be necessary to know more than the fact that a reorder point has been reached at a warehouse. A central planner may need to know no more than that a specific product's stocks at some site have just fallen to the reorder number. Then the appropriate remedy can be taken, whether it is scheduling a production run or disbursing units from a regional warehouse. One of the earliest ways to monitor reorder points and still the simplest is the two-bin system.

The original two-bin system probably actually required two containers that held the same type of item. Today it is more likely that items in a single bin will be tagged to indicate that when the tag is reached it is time to reorder. Reorder points are consistent with a distribution policy that assigns safety stocks, but both just-in-time ordering and reorder point ordering rely on much the same information.

A reorder point is derived from the lead time required to receive an ordered lot of units from its source, such as the production facility or a regional warehouse. Just-in-time ordering also backtracks from anticipated demand due dates to production order release dates by subtracting lead times, so both philosophies of inventory management should work well when they are based on accurate lead times.

A central shipping plan is likely to be strongly based on knowledge of inventories, so this procedure needs to have running inventories and distribution points monitored carefully by the central planning office. Accuracy of information requires that changes in inventories be reported promptly and that actual stock levels be recounted periodically. The ideal situation reports running totals to to a central computer frequently, say daily. The bar coding procedure updates inventory changes in real time.

The most modern bar coding procedure allows for on-line, real-time monitoring of deliveries to customers or distribution points and deliveries from suppliers. The next best inventory tracking is done by accumulating transactions in a microcomputer and then dumping them to a central computer nightly. The old style of keeping track of inventory levels requires flows of pieces of paper that must be recorded after an appreciable lag.

Whatever inventory reporting procedure is used, from manual to fully computerized, counts have a tendency to drift into inaccuracy, and this difficulty can only be countered by performing physical counts of items in warehouses. Probably the best counting method is to check some items every day, concentrating on the important ones. In this way a very important stock may be counted several times a year, while a lesser item may be checked only once.

In any case transactions throughout the distribution system must be recorded and monitored sufficiently quickly and with sufficient accuracy that planning and control systems have a chance to operate effectively. Planning for information flows that will be both timely and accurate is an activity that must be accomplished at an early stage of implementing strategic plans.

If an I.E. is in charge of directing system design and startup he will have an array of questions he will want to be sure he can answer positively.

Questions include:

Can deliveries be assured soon after production runs have been completed?

Can inventories of finished goods, both central and at distribution points be kept sufficiently current in the P&C records?

Can deliveries of shipments be monitored adequately?

Does the order entry/release function keep track of production orders well enough to regularly update anticipated order due dates?

To answer these questions an I.E. would study each constituent operation and verify how it works—leaving nothing to chance. Computerization might not always be required or even recommended. Sometimes information flows can be improved by use of quite simple procedures.

Consider the problem of losing customer orders in the warehouse next to the loading dock. Orders are often lost simply because no one knows where they have been placed. This, in turn, is probably because there is not enough room in the warehouse for all of the orders that accumulate in it. A simple remedy for this problem is to color code an area, perhaps by painting the floor, and then instruct fork lift drivers that orders with tags of given color are always to go to the area painted that color.

If orders begin to accumulate at a faster pace than an assigned space can accommodate, then either the space must be enlarged or shipping must be accelerated. If the space is not large enough, then lower priority items must be displaced, and nothing has higher priority than finished orders ready for shipment. If an A,B,C system is used for sorting warehouse stocks, one would certainly label ready-to-ship lots as A items. Partial lots waiting for completion before shipment would be B items, and items held in inventory to satisfy occasional orders from a regular customer would be C items.

The accessibility of storage locations for items should also correspond to their priority. Some items can be found, idling away in a warehouse, that are rarely requested. These are items someone has hoped would sell but didn't. These do not belong in the warehouse. They should probably be either junked or liquidated.

An I.E. is a good candidate for the person to do the work of analyzing warehouse stocks, procedures, and layout. He can take a hard-headed, hard-hearted look at the way things are done and make objective recommendations aimed at improving flows and efficiency. Objectivity will be needed because the opposition will be considerable. Warehousemen will be found to be attached to the way things have been done for ages. Marketing personnel will be unable to reject from the warehouse an item they forlornly hope someone might buy.

A tag system can be an excellent procedure for tracking A stocks. Every time an A lot is shipped the tag can be routinely handed over to Order Entry/Release, which can in turn enter the information that the lot is in transit. A tag can also have space on it to record the transportation carrier or any other important information.

TRANSPORTATION

Earlier, in Chapter 10, it was suggested that a study of transportation alternatives should be conducted prior to selecting the transportation procedures and modes to be employed. At the strategic level several policy alternatives are possible, such as all transportation being conducted in a single manner. An example might be a company opting to do all shipping only on its own wholly owned trucks.

This would be a broad strategic decision of the magnitude of a facilities decision. However, at the strategic level of planning a transportation study might not produce such clear-cut results. Instead the study might indicate that transportation costs and conditions were constantly too much in flux to enable a simple statement of policy.

In this case the long-term decision might be to construct a routine transportation analysis procedure for constant use in analyzing short-term transportation options. The procedure that comes readily to mind is the use of a common transportation algorithm on a computer.

The usual structure of the model would be:

minimize the sum over i and j of $C_{ij}\, x_{ij}$

subject to:

the sum over i of $x_{ij} \geq d_j$ for all j

the sum over j of $x_{ij} \leq v_i$ for all i

In words this model will seek to minimize the total costs of shipping at least d_j units to each site j via a variety of shipping modes i while not supposing to ship more by each carrier than its current capacity, v_i. The cost coefficients, C_{ij}, are the most difficult numbers in the model to maintain accurately since cost schedules set by shippers tend to vary considerably from time to time and from route to route. Shipping capacities may change a good bit, also.

In order for a computerized transportation model to be useful in selecting shipping modes it will have to be maintained. That is, its database will have to be frequently reviewed and updated. Although an I.E. may initially put the model together and assure that it is up and running, someone else will probably be given the responsibility for its regular running.

Responsibilities for maintenance of strategic systems must be carefully considered at the strategic decision level. No such responsibility should reside at the staff level of the organization because of the natural lack of accountability common to that level. In the instance of the transportation model, whoever is entrusted with its supervision should be under the authority of a line supervisor—say the plant manager or the transportation supervisor.

The role of the I.E. in these operational systems is to design them, construct the information systems requisite for their operation and for their interaction with the operations they should affect, and occasionally monitor them to ensure that they are being used properly. Day to day use of the systems should be conducted by other personnel— preferably someone directly involved in using the system's information. Operational systems should never be left for computer personnel to operate unless they are directly under the authority of line management. The reason is that computer personnel are, like other staff people, insulated from the line consequences of their actions.

SUMMARY

Just as it can be viewed as the front end of an integrated P&C system, so also can distribution be viewed as the primary function of the firm—getting the goods to the marketplace. Once strategies for distribution have been set, the short-term planning and control procedures must be picked. They can range from simplified manual approaches to the use of canned microcomputer software to the installation of a mainframe closed-loop system.

Distribution planning can be based on just-in-time shipments, large safety stocks or even helter-skelter reaction to existing shortages. Computational aids can be of a simulation variety or of a more complex sort, perhaps including some aspects of optimization.

In any case the operation of distribution systems must be planned for well ahead of time. The information structure must be determined early in the implementation of strategic policies. Organizational information flows and lines of authority must also be established so that the systems will work as intended and will be monitored and maintained properly.

The layout and design of such systems seems a natural activity for a strategically aware I.E. to direct. Other related P&C systems also seem to fall into a category of requiring the I.E. expertise and systems viewpoint. Additional major systems of this kind are those that deal with production scheduling and inventory control.

19
INVENTORY PLANNING AND CONTROL

Fundamental inventory policies derive from the strategic plan. A lack of intelligent strategic planning is demonstrated by a production facility that has jumbled inventories, unreliable inventory counts, poor scheduling, and many late orders. Poor inventory management also leads to excessive inventories, which cause low inventory turns and high holding costs.

Inventories should have rational purposes. Raw materials stocks should be maintained to deal with supply uncertainties. The magnitude of the uncertainty and the consequences of stockouts must be weighed against holding costs to set raw material stock levels. Work in process (WIP) inventory is necessary to keep the manufacturing facility running smoothly. WIP levels are tied to the need to alleviate the uncertainties in demands for finished goods and to smooth production flow rates. WIP stocks in effect act as buffer stocks. Similarly, the entire production process may find itself poorly matched to consumer demand, which might lead the company to build finished goods inventories as insurance against missing sales. Inventory policies must fit the needs and uncertainties that motivate them, so an extensive range of inventory policies need to be considered.

INVENTORY POLICIES

Especially at many smaller production facilities, work is directed solely toward filling firm orders. No finished goods inventory at all is kept intentionally. This can be a fully rational policy—especially in the case of expensive, infrequently ordered items. Most larger companies maintain substantial stocks of products ready to ship.

Work in process inventories receive little attention in many manufacturing facilities because attention is concentrated on filling orders. It may be possible that such neglect is rationally supportable, but most I.E.'s would expect that intelligent scrutiny could reduce WIP levels appreciably almost anywhere. A rapid rule-of-thumb for valuation of WIP stocks counts all direct materials currently signed out to WIP as costs plus about half of the other direct costs of production. Thus if a unit required $10 of materials and $6 of direct labor to become a finished good, the average such unit as WIP would be counted as a $13 unit for purposes of quick economic analysis. The major portion of WIP holding costs should be due to the interest cost of investment capital. An I.E. looking over an array of $13 WIP items sees an opportunity for lowering WIP levels, reducing inventory investment, and cutting holding costs.

Raw materials almost surely will require nonzero safety stock levels. One unavoidable reason for this fact is that many needed operational supplies come in indivisable lot sizes that will last for some time. For example, a plant that orders anhydrous sodium hydroxide will probably find it difficult to order a batch smaller than 50 one hundred-pound bags, even though it may use only 8 bags a week. In addition, lead times for stock deliveries and uncertainties about quality, lead time, and delivered lot sizes may motivate larger safety stocks.

Inventory policies have a way of becoming unexamined habit. People tend not to question habitual procedures. At one time it was not too important to question inventory habits because the costs associated with stock levels were not a large percentage of total costs. Such is no longer the case. Inflation since the coming to importance of OPEC has led to drastically larger costs of materials. The accompanying corporate debt, required to finance the purchase of steadily more expensive materials, has led to interest rates that make holding costs doubly prohibitive.

Now inventory limitation has become a topic of broad corporate interest. At the same time technological tools for managing inventories have advanced. Cheaper computing power has encouraged numerous techniques for better tracking and scheduling of inventories. However, the basic aspect of inventory that must be questioned if stock levels are to be controlled is one which almost seems to be unreformed by modern computational techniques. That aspect is the choice of safety stock levels.

SAFETY STOCKS AND LOT SIZES

A typical I.E. cost study will present manufacturing management with rankings that show the relative importance of cost elements. In the case of inventory control it is especially necessary to be able to discern which stock items are important enough to justify stocking costs. The A,B,C ranking procedure is a good start toward cost control.

A items in inventory are those that are particularly important. They may be finished goods prepared for the firm's best customer. They may be items that are so costly as to dominate inventory. They may be pivotal items that must be on hand if production is to continue without untoward downtime. A Pareto principle holds that A inventories must be observed and managed closely.

B inventories are less important. They may be fairly readily obtainable from nearby suppliers. They may be readily replaced by substitute materials. Or they may simply be used in products of only moderate profitability. C items are those that can be dealt with in a relatively cavalier fashion.

An I.E. seeking to rationalize inventory policy is well advised to first ascertain which items should go on the A and B lists. Then he should further investigate those lists to isolate the truly important ones. Just because an item costs a great deal, which automatically puts it on the A list, does not also imply that it is a problem item. Other policies may intervene to make its control trivial.

For instance, a company that makes coal mining machines to customer specs ships every custom-made machine as soon as it is finished. There is, therefore, no finished goods inventory of such machines at all; even though they are extremely expensive products, procedures have naturally reduced holding costs to a minimum.

On the other hand, the same company may also produce a continuous miner of a specific design called LFCM-203. If these machines are made in advance of anticipated orders, their inventory holding costs make them rise to the top of the A list of important stocks.

Just as only the most important stock units should receive primary attention, so also should an I.E's efforts be devoted only to those activities that are on the A list for the entire firm—those activities that are central to profitability. Inventory Planning and Control (P&C) is important to managing the A inventories because of its role in cost control and customer satisfaction.

The closed-loop planning and control systems cited earlier can be viewed as being so interconnected as to submerge the identities of such domains as distribution and inventory control. In a sense this might be so, but not all firms intend to adopt full closed-loop systems, and even those that do will have to adopt inventory policies. And the key to inventory policy is the planning, monitoring, and control of safety stocks.

An I.E. studying existing safety stock levels of A or B items should assume his goal is to reduce them. As a first step he should determine the past history of the item's inventory level. If there has never been a stockout of it, its safety stock is too high. If it has often been depleted, its safety stock may be too low, but not necessarily. If a product's demand is falling, its finished goods safety stock and those associated with its components should be similarly reduced.

This is the dynamic feature of inventory P&C. While strategic policy may be fixed—say, ''allow stockouts only 5 percent of the time'—current implementation of the policy must be based on current conditions. Thus, safety stock levels calculated at one time must be revised as the conditions that gave rise to them change. As a case in point, consider EOQ lot-sizing policy.

EOQ (economic order quantity) is a procedure every I.E. is taught several times in his collegiate curriculum. Under its elementary assumptions, a set production order release for a fixed lot size of units is intitiated when stocks are exactly depleted. No safety stock is at first associated with the EOQ order policy. Only after uncertainties of demand levels and of production lead time are factored in do the textbooks tack on safety stock to the EOQ lot size. The usual combined formula is:

$$\text{Lot Size} = \text{EOQ} + SS$$

where EOQ is the square root of $2PD/H$

where P = order (or setup) cost

H = annual inventory holding cost per unit

D = annual demand

SS = safety stock

The assumptions this formula is based upon are:

Order costs, including setup, are large.
Holding costs are also large.
Long-run demand is predictable and large relative to lot size.
The consequences of running out of stock are costly.

Perhaps it is because this type of formula is so pervasively taught that it is widely implemented in production facilities to which it does not apply. In fact, it is not uncommon to find the formula asking for unrealistically large lot sizes. A study of the formula reveals some pitfalls surrounding its use.

What if order (setup) costs are quite small? Then the EOQ should be small, approximately equal to safety stock alone. However, setup costs are often overestimated to the point that the EOQ formula produces an especially large lot size, which, instead of balancing off order costs against holding costs, incurs extra holding costs. An I.E. faced with the use of such a formula and armed with the information that the product has rarely experienced stock-out would be well advised to study the origins of the numbers being used. Most companies over recent years have seen their setup costs decline and their holding costs rise to such an extent that their EOQ formulas should be abandoned.

Similarly, safety stock calculations should be checked. Logically, safety stock should be motivated only by the desire to avoid being out of stock. If the consequences of being so are negligible, safety stock can be ignored. If safety stock seems to be called for, its level should be revised dynamically as costs and demands dictate.

One must remember that the actual safety stock being employed is the average number of units in stock. Production lot sizes that are too large and safety stocks that are too large can both create expensive, excessive inventory. In modern, integrated manufacturing facilities most products can probably be governed in a just-in-time fashion so that safety stocks need be little more than expected depletion during production lead time.

A better approach to evaluating inventories than "EOQ plus safety stock" is simulation. In this approach past data are used to construct a profile of demand fluctuations. The derived profile, adjusted for trends, is used to project product demands, which are then compared with planned production levels. First, simulated stockouts resulting from the planned production are observed and compared with stockout criteria; then production levels are adjusted up or down as the simulation results suggest; these steps are performed iteratively until a production policy is reached that satisfies the selection criteria. A commonly used rule is to plan for production output to be as small as possible while allowing stockouts no more than X percent of the time.

Unless setups are unusually expensive it is unlikely that setup costs need be a major consideration in choosing lot sizes and safety stock levels. Lot sizes in some cases may be fairly inflexible. For example, the firing of bricks in a kiln is ordinarily performed in fixed lot sizes with similar intervals between lots due to the characteristics of the kilns. In such a case the only real question to ask about safety stock is how many lots to schedule.

The calculation of dynamic lot sizes and floating safety stock levels assumes the availability of accurate, current data describing sales, good predictions of demands, and accurate perpetual inventory counts. If these are to be good assumptions, they require well-developed information systems. In fact, the ability to plan stock levels of finished goods rests on the accuracy of information all the way back to the beginning of the production chain—starting with deliveries from suppliers.

SUPPLIER PLANNING AND CONTROL

Nothing disrupts a production plan more surely than missing raw materials and subassemblies. If a company's ordering procedures are sound, then the dominant source of this problem is suppliers. In the absence of perfectly predictable deliveries, the stockroom may require raw material safety stocks.

The same reasoning as was applied to finished goods can also be applied to raw materials. Lead times must be ascertained with as much accuracy as possible. An I.E., casting about for appropriate stock levels, will be interested in studying the arrival dates and delivered quantities by supplier and correlating them with their orders from Purchasing. He would also be very interested in the results of acceptance sampling of supplies from each major supplier. A,B,C categorization should again be used to avoid wasting time studying minor suppliers.

As was mentioned much earlier in this book, data used to plan production begins with reasonably accurate descriptions of the reliability of raw material and subassembly stocks. Standards for material requirements throughout the production process begin with an understanding of the quality of incoming supplies. Such an understanding can only be developed by persistent observation of those incoming supplies.

There are three main elements that must be monitored—a fact that should shape the system used to monitor stockroom supplies.

1. Lead time between purchase order and delivery of order must be monitored over time.
2. Accuracy of deliveries, such as the amount ordered vs. the amount delivered, must be tracked.
3. Quality of delivered quantities as determined by acceptance sampling must be ascertained.

Only when the variability of lead times and quantities delivered are known can safety levels of important stocks be determined. Also, only when this basic information is known can intelligent decisions as to selection of suppliers be made.

Stockroom monitoring is important to do well, but it is something that is not often done with great accuracy. Just as variations in the reliability of suppliers contributes to the need for safety stocks, so does variability in the preciseness of stockroom management. The nadir of raw material control is found in organizations that have no idea what is in stock nor in what quantities. The authors have encountered stockrooms that suffered from this very lack of knowledge.

At the other end of the management spectrum are companies that know almost to the item what is in stock and in what quantity. Such companies have spent time and effort developing P&C systems that work and have enforced the discipline required to make them work. It makes no difference whether a system is computerized or just meticulously maintained manually, stockroom management requires careful attention to detail.

It is interesting to note that if finished goods safety stocks and lot sizes are well understood and if raw material safety stocks and ordering procedures are well understood, then the key to a successful operation lies almost entirely with disciplined production planning and control.

INVENTORY CONSEQUENCES OF PRODUCTION PLANNING AND CONTROL

A simplistic view of P&C systems would hold that they consist of nothing but scheduling and inventory control. An even simpler view holds that inventory control is merely one facet of scheduling. This view asserts that the scheduling of production orders and purchase orders is the essence of inventory control.

However, it is clear that inventory policy is developed quite apart from scheduling techniques. In fact, the choice of inventory policy dictates scheduling procedures. Just-in-time scheduling derives from a determination to avoid inventories of all kinds. A decision to eliminate buffer stocks in an assembly line dictates the way the line is scheduled and how work stations are designed. A decision to limit stocks of finished goods affects distribution scheduling.

The upshot is that inventory control and scheduling are two interrelated, but to some extent independent, aspects of operational P&C systems. At the strategic planning level, policies to guide both tools must be adopted, along with the day-to-day format for implementing inventory and scheduling decisions.

Consider the front end of a typical scheduling algorithm as a demonstration of the intertwined effects of scheduling and inventory. Demands for products are developed from a distributed requirements planning algorithm, forecasts, and firm orders. The result is a preliminary master schedule for products. From the management perspective a master schedule is just that—a list of quantities and due dates that production is to satisfy.

From the production side and from the broader perspective of operational management, a master schedule is the entire set of production schedules, purchasing schedules, and timings of inventory levels. A minor example is shown below to demonstrate this viewpoint. A single product has quantities and due dates as shown below on the display. In the lower parts of the table the production and inventory consequences of the due date schedule are shown. Note that the product's manufacturing lead time is two weeks.

	Week 1	Week 2	Week 3	Week 4	Week 5
Order Due Dates	10		30		10
Planned Order Receipts	10				
Inventory On Hand 10	10	10	0	0	0
Production Order Release	20		10		

The derived schedule satisfies a natural scheduling criterion— the just-in-time production of required units of production. However, inventory policy has not yet been considered. Order releases are scheduled for the first week of the planning period and for the third, but inventory has been allowed to drop to zero after the second week.

The second part of the front end of the P&C algorithm used for scheduling is probably the inventory control module. In this case, let us assume that a minimum inventory (safety stock) level for the product in question has been set at 10 units. Then, given the master schedule as presented above, the inventory control algorithm would revise the last two lines as follows.

	Week 1	Week 2	Week 3	Week 4	Week 5
Inventory On Hand 10	10	10	10	10	10
Production Order Release	30		10		

This small example for only one product demonstrates the separateness of the logic associated with inventory control and scheduling. In particular it points up the basic conflict between safety stocks and just-in-time scheduling. At a more detailed level this conflict is revealed in later modules of the production P&C system—in the MRP segment.

MRP (material requirements planning) simply projects the steps of producing an item backward in time and determines a production schedule for each constitutent manufacturing step. At this level of detail both inventory policy and lot sizing decisions affect inventories.

Say that the production of 10 units of a final product require that 40 small parts of a given type be included in its assembly. Assume further that the small parts must be started into production a week before they will be needed for final assembly. The number needed and the production order release date for the small parts is now known, but two other questions must be answered that impact inventory levels. These questions are "What lot sizes are required" and "What safety stocks must be kept?"

The answers to these questions determine both the production run sizes for the small parts and the resulting inventories. All through the planning process the problem is the same—schedules cannot be established without reference to both production and inventory policies governing lot sizing and safety stocks.

The only case in which scheduling can be conducted without cross-referencing inventory considerations is when the plant operates in an exact just-in-time manner. Few American facilities have operated this way in the past, but there is a current trend toward this behavior. The ability to reduce design and setup times provided by modern technologies has also led to a decline in the importance of economies of scale, which in turn reduces lot sizes and lead times. However, the basic information required to operate is still the same.

Accurate lead times must be known to lay efficient plans. This means that fairly reliable averages and standard deviations of important quantities should be known. It is said that the Japanese kanban mode of scheduling, which eschews inventories, allows no more than a 5 percent deviation from expected times in assigning job time standards. This is considerably smaller than American firms permit and is only viable due to the highly repetitive, tightly monitored design of kanban operations.

Even though American companies may add a larger safety margin to mean times for jobs, the standards still need to be known and reviewed. So must material standards, etc. These are the classical domain of the I.E., but the modern I.E. should extend his sphere of influence to cover the interface between inventories and schedules.

The I.E. should be privy to the strategic considerations upon which inventory scheduling decisions are to be based. If the policies in use are based upon mere unquestioned habit, such as is often the case when EOQ is the inventory policy, he should bring them to question. If automated procedures, such as closed loop P&C systems, are to be installed he should be prepared to ask whether they fill a real need and whether they will be cost-effective.

Given the common current assumption that inventories are very important to the company's profitability and its ability to compete, they are surely of prime importance and deserving of the I.E.'s attention—from the development of the strategic underpinnings of inventory policies to their operational implementation.

SUMMARY

Inventories have become simultaneously more costly to hold and less costly to govern. However, accurate governance requires an information system based upon accurate and regularly updated data. In particular, it is necessary to have good counts of stock levels, reasonably current information on stock depletions, good standards throughout the acquisition and production chain, and a firm understanding of costs.

In addition, the I.E. who seeks to put inventory control on a rational basis must understand the strategic policies that have been adopted. Only when both the bottom-up operational understanding and the top-down strategic view converge can rationality be expected to prevail.

20 SCHEDULING

Scheduling is arguably the most important production function. Important determinants of a manufacturing company's health depend on scheduling—when it's done well, the health is good, and when scheduling is done poorly, the company does poorly. The most visible measures of corporate performance, revenues and costs, are both affected by the timing of production.

At the end of the entire sequence of corporate activities is the customer, expecting service. And service usually means deliveries of products on time and in the right quantities. In the production steps themselves, well-organized scheduling reduces scrap, rework, expediting, in-process inventories, and raw material inventories. Yet, in spite of its importance, scheduling often receives scant attention at the strategic planning level of corporate activity.

SCHEDULING AND STRATEGY

As has been suggested for all operational systems discussed so far, the short-term abilities of scheduling systems must be built into the company operational structure as part of its long-term preparations. Otherwise the common frailties of scheduling procedures can impair the firm's ability to achieve its strategic goals.

Management's most grievous error with respect to scheduling is its allowing diverse personnel to meddle with properly prepared schedules. Hard as it may be to believe, many a company will spend princely sums acquiring, tailoring, and implementing a complex computational scheduling system only to permit intermittent contravening of its work. The most common intervention is permitted on the part of Marketing.

Salesmen naturally want to see their customers, who represent high commissions for themselves, receive favored treatment. Salesmen would cheerfully change the production schedule—halting everything for a favored customer's rush order. Although such a change might not lie outside the bounds of scheduling normalcy, it is probably ill-advised. Worse, the very fact that it can happen shows a lack of strategic insight into operational matters.

This is a point that a strategically aware I.E. can bring to the conference table at policy adoption time. He can stress the fact that scientifically derived schedules that are specifically chosen to advance the firm toward the achievement of its objectives should not be tampered with on the spur of the moment. What is meant by a scientific approach to scheduling will be discussed below.

A second dominant error in scheduling, and one that is also surprisingly common, is loading plant facilities so that input loadings equal theoretically maximum flow rates. In this case the reasonably predicted flow capacities for all processing steps are assembled and production releases are scheduled to match the capacities exactly. The error is a clear violation of observance of the variation effect, discussed in Chapter 5. Since processing capacity cannot be stored, the times when the work stations drop below their maximum operating rates result in the accumulation of needless in-process inventory and delayed jobs. Jobs should be scheduled according to past observations of average flows, not maximums.

The third dominant error in scheduling jobs is ignoring interaction effects between jobs. For instance, a set of jobs that use a certain capacity-constrained machine may all be scheduled on the same day, assuring that they will not be finished. Another common interaction effect is found in the scheduling of jobs with sequence-dependent setup times. There is often little effort devoted to trying to select the quickest sequence.

The fourth prominent scheduling error is not providing for schedule revisions. In olden days before reliable computer programs could perform the bulk of the scheduling work, it was not uncommon for a company to persist in using a schedule that had long been clearly violated. It simply took too much time to recalculate. The likely reason for this error cropping up in modern manufacturing facilities is that it was not anticipated at the strategic level.

The reader might have come to wonder by this point in the book why the authors lay so much blame on the strategic planning process when operations go awry. Especially in the case of scheduling the explanation is easy to display. In the press of day-to-day activities all operational personnel are devoted to fulfilling the schedules at hand. They have no time to investigate diverse alternatives nor to research better techniques for timing and sequencing. It is only during the relatively leisurely process of long-term planning that such questions can be addressed effectively.

Thus, poor scheduling in the short run means there was poor planning for the long run. One aspect of long-term policy making that must be stressed is that someone in the firm must be able to find time to study operational procedures like scheduling to determine whether they are functioning according to plan.

As has been stressed, such a person should probably be an I.E. who has been given just such responsibility.

Only when responsibility, time to do the job, and money to support the work have been provided will the job be done. I.E.'s naturally obtain a systems viewpoint of production and realize that effective operations must be coordinated, whereas personnel in specific departments tend to see scheduling as a problem peculiar to their limited area. Thus, a wise strategic move for management to make at planning time would be to appoint an I.E. as an overseer of the operational systems that will be adopted.

This suggestion is only one part of a general admonition about operating systems. Too often management elects to employ procedures, such as computerization of some function, but neglects to budget the resources to maintain the procedure and the systems involved. Resources required to keep any system going include personnel, current expenses, maintenance resources, and time for monitoring, review and evaluation.

Consider the special case of scheduling. If jobs are badly timed the repercussions can be costly, so scheduling systems should be monitored and reviewed almost constantly. This means someone must be in charge of the effort, and he must have authority to take preemptive or corrective action. Funds must be readily available to support maintenance of both hardware and software. The personnel appointed to overseeing important scheduling functions must be provided with the time to do the work. All of these resources must be planned and provided for at the strategic level.

At the strategic level of planning the net implication of the above discussion is that planners must adopt definite corporate guidelines for scheduling along with a systematic approach to maintaining the chosen scheduling process, and management must commit to the effort required to produce and implement good schedules.

SCHEDULING GUIDELINES

At the top of the list of strategic scheduling decisions is the choice of scheduling procedure. This choice is increasingly becoming viewed as a choice of hardware and software. This is a rather limited view because no computer system will run itself. Nor will it maintain itself, nor will it develop its initial database. It will not update its data, either, except in the case of fully automated control systems.

Let us assume that your company has opted to buy Gigocorp's closed-loop manufacturing resources management system, GIGU-109. Included in this system is a set of procedures that effectively define your scheduling methods. The system must next be fitted to your operating environment, which means it must be tailored to meet your requirements. At the same time your procedures must all be changed to satisfy GIGU's requirements. Someone must be in charge of all this installation and adaptation. The responsibility should not be entrusted to any of the corporate subdisciplines such as Accounting, Marketing, Production, or Computing—especially not Computing.

The basic management conviction that supports the adoption of integrated systems is that their use objectifies decisions and forces everyone to adhere to the central philosophy the system implicitly implements. However, if some subgroup gains control of the system, it will surely use the new tool to its own advantage, regardless of the effects on the corporate plan. This is human nature. This is why the company needs someone who is specifically devoted to keeping the company as a whole on track and who has the systems viewpoint needed for the job. A strategically aware I.E. would be ideal.

Granted, one's company might decide to adopt a simple scheduling system without all the bells and whistles that vendors promote. Alternatively, one might have a computer program written in-house and run on a microcomputer. One might even maintain the time-honored procedure of having an old hand at scheduling keep doing it. But in any case the arguments above prohibiting the dominance of the individual departments are still valid.

Scheduling should be directed by corporate policy. That is, all short-term aspects of procedures are to be tested against how they advance corporate objectives. It is the job of upper management to understand what policy alternatives exist, what they mean, and which are to be in force. Such technically trained personnel as I.E.'s may provide the bottom-up analyses that will support the decision, but corporate strategy choices ultimately must be made by those with the final authority.

Just as the strategic plan must be chosen for the good of the entire firm, so must individual procedures be adopted to conform to the overall plan. Thus, no single department can be allowed to have hegemony over any activities not specifically in its domain. However, conflicts arise over the governance of scheduling. Marketing naturally views scheduling as its domain because final sales, it believes, represent the sole goal of the company. Production sees scheduling as solely a production activity because a manufacturing firm exists to transform resources into final goods via the production function. A natural appeal of the large-scale integrated P&C systems is their apparent ability to resolve such conflicts.

Another policy issue central to scheduling involves the setting of priorities. In particular, the question that constantly arises is how to resolve scheduling conflicts. Such conflicts are ordinarily resolved by the use of priority rules. Priorities descend from the strategic plan and ascend from operational realities.

At the top end, the policy may be to emphasize customer service, so the scheduling criterion is to avoid tardiness of customer deliveries. This consideration may also impose the onus of safety stock maintenance on production managers. A completely different priority might be the minimization of safety stocks, even the absolute limitation of inventories. Another priority rule altogether might be derived from a decision to employ a level loading policy on plant operations. Yet another priority might be achieving maximum flow—the maximum utilization of production capacity. One more priority rule is worth mentioning—incurring low operating costs per unit.

Both level loading and stocking-to-inventory policies force the schedule to be fleshed out with inventory stocking runs whenever slack capacity remains after scheduling forecasted orders. This policy increases total costs by adding significantly to inventory holding costs. However, inventory building can also be a consequence of trying to hold costs per unit of production down by planning long runs.

If minimizing tardiness of orders is the primary scheduling goal, it is hard to say how it might be satisfied simultneous with maintaining high inventory levels. If only a few products are being made, then high inventories would almost preclude stockouts. However, if many products are candidates for sale, long production runs may seriously interfere with the ability to find time for specific orders.

Conflicting criteria create the need for methods of conflict resolution, and one such method is the use of priority rules. To demonstrate the need for specific priority rules consider a simple example. Product A has been assigned a production lot size of 100 units, chosen to incur a low setup cost per unit while not overstocking inventory. Product A has a minimum inventory level of 25 units, based on average inventory depletion during the lead time needed to produce another lot. Lead time from production order release to lot completion is two weeks. Let us assume the following current data.

Table 20.1. Worksheet for Production Scheduling.

DUE DATES	WEEK 1	WEEK 2	WEEK 3	WEEK 4
Order Size		80		50
Starting/Revised Inventory	100/100	100/20	20/120	0/170
Requirements	0	100	0	100
Planned Order Release	100	100		
Planned Order Receipts			100	100

In this small example the first cut through the schedule is an attempt to satisfy due dates, but it is found that doing so reduces safety stocks below minimum, so requirements of 100 units of product A are found to be needed in weeks 2 and 4. However, the two-week lead time for production has forced the schedule to slip a week for the first lot's order release. As a result, actual ending inventory in week two drops to 20 units, i.e., below safety stock. The shortfall is made up as soon as possible in week three.

Another aspect of this schedule that bears attention is the actual inventory size. Average actual end-of-week inventories are 102.5, which is large relative to the safety stock of 25. The reason for the seeming overproduction is the choice of lot size. It is interesting to study the scheduling procedure used in the example to determine the true priorities invoked.

The first priority is that due dates are to be delayed only as a last resort. The second priority is that production runs are to be scheduled as late as possible and only when required. The third priority is that lot sizes are not to be violated. The last priority is that safety stock requirements can be violated rather than delay due dates.

Other priorities could have been adopted. For instance, a rational policy would restrict lot sizes to that quantity needed to fill orders or refill safety stock. Whether this policy could be adopted would depend upon what needs or constraints cause lot sizes to be selected.

The point to this little exercise is that priorities will dictate schedules. They in turn dictate inventories. It is therefore imperative that priorities not be allowed to simply evolve without rational direction. Every firm must decide what profitability criteria should guide its inventory policy. These criteria in turn suggest priority rules. Homogenized approaches will probably not work; the rules will need to fit their working environment.

Fundamental strategies must be adopted first. If manufacturing is to be viewed as a transformation process—changing materials and resources into goods and services—then it is driven by some specific motivation. Probably the motivation is perceived demand. Thus, management must develop firm strategic guidelines for how demand is to be derived and what lines of authority will be allowed to affect estimates.

Next, priorities must be adopted that facilitate the satisfying of product demands. In particular, it must be determined how to resolve the conflict caused by two orders needing access to scarce production capacity. A common rule of thumb is to schedule the products by the shortest weighted processing time procedure in which an order's processing time is divided by the weight given that order, and then the product with the smaller ratio is made first.

There are other common priority rules. For instance, consider a company that makes most of its money from its sales to a small group of customers—the A group. (We would naturally assume that other customers are either of the B or C designation.) The rule in this case is that in a conflict over capacity utilization an A always precedes any B or C product. Within a class of products perhaps a version of the shortest processing time rule could prevail.

For example, let us look at the short list of current orders a plant is trying to schedule. The order names are shown along with their capacity requirements and processing times in weeks.

Order	21	30	43	52	67
Capacity Requirement (%)	30	35	25	20	30
Processing Time (Wks)	8	10	7	7	6

The table shows that if all the orders were put into the system at once 140 percent of capacity would be required. If the shortest weighted processing time were used to prioritize orders, the ratios could be derived as follows, using inventory costs as weights.

Order	21	30	43	52	67
Inventory Cost in $1000/order	6	8	11	12	6
Shortest Weighted Processing Time (Wks)	1.33	1.25	0.64	0.58	1.0

The rationale of reducing an order's processing time inversely as its value in inventory is motivated by the desire to limit inventory. Thus, the higher inventory dollar values would be shipped as soon as possible. The resulting product priorities are: 52—43—67—30— 21. This sequence allows the first three orders to be run now, but 30 and 21 would have to wait until capacity became available.

An I.E. studying scheduling procedures in his company will be interested in which rules are actually being used, how they were derived, and how their performance can be evaluated. Only after policies and their attendant priorities are intelligently determined and installed can the scheduling system be expected to work right.

SCHEDULING SYSTEMS

In the past, scheduling, if it was not entirely of the "seat of the pants" variety, consisted of arraying jobs on a Gantt chart display. Some companies kept large metal boards on which they placed magnetic cardholders containing cards with lot numbers and lot descriptions on them. The cardholders could be scooted around on the boards by an expert scheduler until they were aligned into a reasonable sequence; then the layout of jobs on the boards would become the schedule. Revisions were accomplished in much the same way.

For many manufacturing firms this procedure is still viable. In many smaller shops there is little complexity to scheduling. Other firms operate in such a repetitive fashion that schedules can be prepared with little variation for periods as long as several months at a time. However, most manufacturing operations can benefit greatly from a more systematic scheduling system, which may or may not require computerization of basic production data.

Whatever system is adopted, the operational goals of good scheduling are to produce work on time, avoid overstocking of inventories, and maintain smooth operations. The system must also be capable of developing and maintaining the data required and updating the priorities demanded by the strategic plan. A good system must anticipate problems such as material shortfalls and missed due dates. It should also be capable of ready revision.

There is a heirarchy of scheduling methods for manufacturing facilities according to the difficulties involved and the portion of the work that is to be computerized. At the simplest is the single, human scheduler who does any necessary calculations with a trusty calculator. Someone will probably have to help with purchasing and inventory counting, but a small plant can get by using this old approach.

One step up in complexity is scheduling that requires a good bit of calculation, but its primary requirement is human intuition. This situation is ideal for the inclusion of a decision support system (DSS). In this case an experienced scheduler sits at a terminal and enters what he believes to be a good sequence. Then the computer analyzes the plan numerically, possibly writing out a detailed schedule and listing of capacity utilizations based on the data submitted by the person. The human half of the team then surveys the output and makes revisions as necessary. When he is at last satisfied with the detailed schedule, he interrupts the iterations and adopts the new itinerary.

The DSS approach is attractive in many instances because it permits considerable discretion on the part of the scheduler, who may be dealing with many unprogrammable constraints. A DSS can be fairly inexpensive—another point in its favor. It can probably be run on a microcomputer, possibly using a simple canned program such as a spreadsheet analysis software package. DSS is also a good solution to special-purpose scheduling problems, such as creating a detailed departmental schedule from a rather broad schedule sent down from above.

The next step up in the hierarchy of complications is a large-scale scheduling system that is primarily computational. If there are large amounts of computation required to prepare sequences and schedules along with a large amount of data to be regularly updated, DSS is not a good solution. However, a predominantly automated procedure imposes a number of constraints on how things are to be done.

For instance, unlike the DSS approach, a fully automated scheduling algorithm must be programmed to satisfy all constraints, policies, and priorities. Thus no unprogrammable constraints are permitted. Data needs of the algorithm must be maintained on file all the time. All pertinent scheduling data must be assiduously delivered to it according to a thorough plan for doing so. These characteristics of computerized scheduling impose discipline on the organization that invokes it. Discipline consists of planning the procedure to work, teaching personnel to perform with the computer in the right way, maintaining all facets of the man-machine system, and adhering to the operational procedures upon which the computer algorithm is based.

Some workers complain that working with such a system turns them into little more than clerks who chase down data for the machine. On the other hand, a disciplined organization implementing tight schedules can reduce WIP inventories almost immediately and keep them down while it simultaneously improves its on-time delivery record.

Computerized schedules are almost indispensible at many facilities, such as large-scale job shops. This kind of operation needs a thorough method for correlating orders, inventories, purchases, and production schedules, or else it invites chaos. A computerized system can help manage these functions. Yet there is still one more level in the hierarchy of scheduling complexity that takes computerization one step higher.

Many companies are now considering installing a large computer system known as a closed-loop production planning and control (P&C) system, also known as an integrated manufacturing system (IMS) and as a manufacturing requirements planning (MRP II) system. The latter name is chosen to identify this system's origins with the original material requirements planning (MRP) systems.

Closed-loop systems in outline can virtually run an entire production facility. However, few companies claim to have installed the full set of modules in a complete system. Total integration requires tying accounting, finance, personnel, marketing, and production functions together. The number of firms considering such an installation is growing, however.

Whatever type of system is chosen by management for scheduling will have the effect of defining how operations are to be run. The more computerized the system, the greater the discipline imposed on operations. The great benefits that can be gained from a well-regulated operation are lower costs and better performance, but some flexibility may be lost. In consequence, the selection of the degree of computerization so colors the eventual operating philosophy that it must be a long-term strategic decision.

Similarly, the short-term characteristics of the selected scheduling system must be delineated at the strategic level. This is where I.E. assistance is valuable; someone must actually study operations and determine what levels of complexity operations demonstrate. It is often easy to tell whether a change is needed in scheduling procedures. The way work is progressing on the shop floor can provide conclusive evidence of scheduling needs to an alert I.E.

If scheduling appears to need help, the question is which level of the computerization hierarchy does it require. Unless the system in use now is providing the information needed by a computer system, it cannot be computerized successfully. An I.E. can determine whether the data requirements of an automated procedure can be met. Consider, for example, the kinds of information that must be routinely provided for scheduling systems to work.

MASTER SCHEDULING

Master scheduling consists of checking the product due dates against inventories, lead times, lot sizes, and anticipated order completions to determine a tentative production schedule. If the derived schedule can be fitted to the production plan so that all due dates can be met, the initial master schedule stands as is. Otherwise, due dates are postponed until the production schedule can satisfy them.

For every product this algorithm requires:

a current inventory count
an in-process production schedule
product lead time
production lot size criteria

In addition, if the product has a safety stock criterion, it must be known.

If the master scheduling algorithm is to work automatically without a large amount of human intervention, it must also be equipped with the logic to resolve conflicts. It must be able to choose between delaying due dates and dipping below safety stock. It must be able to be updated when some of the original data, such as anticipated order receipts, change. It must be able to determine whether lot size limits can be violated.

An alternative to fully automating master scheduling is to simply have the algorithm print out the conflicts. Then a production manager can decide what to do. If the computer predicts an order will be finished late, the manager may decide to expedite the job by assigning extra resources to it. A little expediting is unavoidable, but a lot of it means the scheduling system does not work.

The I.E. studying scheduling at a facility must determine what kind of system will work. This information should be the basis for the strategic choice of operating system. An I.E. can readily determine what information is readily accessible in the production environment and whether it will be sufficient to operate a specific scheduling system. To do this work, he will also have to become familiar with candidate systems.

The usual mistake that management makes in this regard is to assume that only a computer programmer can possibly understand a closed-loop system.

Thus, a programmer is dispatched to evaluate candidate systems, which he does without knowledge of the operating environment and without interest in costs. The error is clear, as is its remedy. An I.E. should do the operational and cost analysis.

MRP

The main component of large-scale manufacturing scheduling systems is the MRP module. It explodes a master schedule into components, subassemblies, and fabrication steps, and projects needs for purchases of parts and raw materials. Its main value is that by working backward from due dates, it can anticipate shortages and delays.

The data required for an MRP system to work are truly daunting, including for every product, production step, and material a usable lead time for performance or acquisition. Every product must have a full and up-to-date bill of materials. To tie this information in with Purchasing, every item must have a reference number for purchasing and inventory control. Inventories of every item must be kept current. A manager considering installing such computerized systems had best be warned that the job will require the gathering, storing, and maintaining of thousands of pieces of data. An I.E. can easily explain this point while he is reporting on what kind of scheduling system should be provided for in the strategic plan.

The kind of system selected should solve three of the most common scheduling errors if run properly. It should prohibit unwarranted schedule interference and overbooking of facilities, and it should provide for revising the schedule. The first is achieved by management's adoption of definite policies with regard to schedule interference and the operation of the scheduling system. The second is achieved by installing intelligently derived time standards in the scheduling algorithms. Schedule revisions can easily be accomplished by any level of computerized scheduling procedures.

The I.E.'s role in these systems should be one of evaluation prior to selection, development of tailoring and implementation strategy, and occasional monitoring thereafter. He should not be involved in its day-to-day operation nor in laboriously maintaining standards. However, one mode of scheduling might require his concerted attention.

PROJECT SCHEDULING

When a company contemplates a large, lengthy project the natural questions are how long will it take and how much will it cost. An I.E. is a prime candidate for coordinating the work that will answer these questions and lay out a project schedule.

The probable technique used to schedule a project is some version of the PERT (Project Evaluation and Review Techniques) procedures that I.E.'s learn in school. However, project scheduling is not a job for an I.E. alone. As coordinator of the effort the I.E. needs planners familiar with the type of project under consideration. The crew might include architects, cost engineers, legal people, and representatives of any other relevant disciplines. The product of the project evaluation work should be a schedule of the project along with time and cost estimates for the discrete activities selected to constitute the whole, but the most important part of this schedule is its capacity for review and revision.

While a PERT network of activities that make up a project can produce a Gantt chart of the implicit schedule, replete with slack times and critical paths, the full needs of management are more complex. Management needs to be able to track the activities and always be able to discern whether a job is on schedule and whether the budget is on schedule. In addition, management needs to know what its contingency options are as fully as possible. The information system attendant to satisfying all these needs is quite complex and difficult to design.

To date there does not appear to be any fully successful way to monitor, review, and repair long-term project schedules, so the coordinating I.E. must be content to simply do the best he can. However, a truly efficient system for project planning, costing, scheduling, tracking, revising, and controlling would be a great boon to the world. Perhaps some industrious I.E. will devise one.

SUMMARY

Scheduling needs vary widely, as do tools for doing the job. However, the main role for I.E.'s in this area appear to be in evaluating, selecting, implementing, and monitoring scheduling procedures. The last great scheduling sin, not accounting for job interactions, has not been dealt with in this chapter because the topic crosses over from scheduling into another large area for I.E. involvement—capacity utilization.

21
FACILITIES UTILIZATION

While scheduling is considered by many analysts to be the source of many problems in manufacturing firms, the discord associated with poor scheduling often has its roots in misuse of facilities. Whether management does not understand facility utilization concepts because of a lack of on-the-floor experience or because of more basic misapprehension is difficult to decide. Perhaps both factors contribute to improper facilities utilization.

As was mentioned in the prior chapter, one of the most common scheduling errors is overloading facilities. This invariably leads to large in-process inventories and an abundance of missed due dates. At the root of this error seems to be a misunderstanding of product flows. Other causes of poor facilities utilization appear to spring from a lack of feel for how interactions among various production entities affect schedules. This error rests on the belief that individual products are independent of each other.

It may be that management simply does not understand its production capacities. Or it may be a lack of knowledge of aggregate planning. Whatever the reasons, misuse of facilities is very common. It is the job of the I.E. to bring rationality to planning that ultimately dictates how plants, machines, and people are to be employed.

CAPACITY LAYOUT

It is to be expected that a new facility will be laid out so as to rationally perform its function, but older facilities commonly violate this criterion. As plants grow or diversify over time it is usual for them to tack on activities without ever reevaluating the resulting implicit production flows.

It is not unusual to find a foundry that moves hot metal over a finishing area, thus requiring the finishing workers to move out of the way until the hot stuff has passed. Other plants may crowd new machines into the old layout until congestion prohibits the machines from all running at the same time. Confusion often results from unplanned growth and the resultant jerry rigging.

An I.E. might find it instructive at an older facility to simply sit away from the action and observe traffic. It can be found in many cases that the traffic crisscrosses itself many times, creating confused patterns of crisscrossing people and forklifts and material flows.

This problem must be resolved before facilities can be made to operate rationally. Once attention has been brought to bear on confused flows some progress may be possible in correcting them. Certainly no progress will be made without scrutiny. And only after corrections have been made does it make sense to try to make scientific capacity utilization analyses.

However, confused flows should not have been allowed to occur originally. Neither should misuse of capacity occur. Toward avoiding misuse a plant should perform a thorough facility loading analysis at the time of plant startup.

FACILITY LOADING

No I.E. would knowingly schedule plant operations by booking any work station at 100 percent of its nominal capacity. He would obtain machine standards based upon how the machine should operate on the average and prepare the machine loading schedule from them. The use of maximum operating rates for booking purposes ignores the fact that a great deal of the elapsed time in most production processes is not available for production.

Machine loadings should be based upon time standards that implicitly include the effects of down time, idle time, rejects, rework, and time to smooth operations. The variation effect applies to the scheduling of assembly lines, machines, work stations, and entire facilities. If one merely takes the actual time a machine is used directly to work on a product as a standard for machine loading, the variation effect will ensure that the work-in-process bin at that machine will have to be very large.

The variation effect prohibits closely matching flow rates. A theoretical result from queueing theory shows that if the input flow rate to a process is close to the average processing rate there will be a large queue of jobs awaiting processing. The closer the two rates match, the longer the queue.

In fact, the capacity of most machines or work stations is often irrelevant to the proper loading of a plant. Ordinarily there is a small number of steps that constitute bottlenecks in the production process. A bottleneck could be caused by material handling capacity, or the capacity of a particular work station, or a quality control center's inspection pace. Once a bottleneck is discovered, it becomes clear that the bottleneck in the process cannot be overloaded, so it determines the overall utilization of the entire sequence of steps including it.

I.E.'s should be well versed in the awareness of bottlenecks, since they are trained to conduct the objective analysis that reveals them. Aggregate planning analysis, performed with correct standards and revenue figures can simultaneously reveal the best sources of profitability and the bottlenecks that constrain output.

AGGREGATE PLANNING

Proper scheduling of manufacturing operations depends on sequencing and capacity utilization. Although any specific schedule must be analyzed according to current product demands, it is a good idea for management to understand the effects of product interactions and capacity conflicts in maximizing profits. Toward this end the firm needs access to a plant loading model that can examine "what-if" questions about capacity planning.

The objective of studying the plant with a computer model is not necessarily to lay out production schedules; the main goal may be to discover the important scheduling considerations. A person trained in understanding and interpreting computer model output can run such a model under a range of operating assumptions to discover the important features of capacity utilization that must be built into the planning of schedules at any specific facility.

For instance, running a model to discover the optimal product mix in a given situation may reveal that certain products tend to use up all of a capacity bottleneck so rapidly that more profitable products are squeezed out. Larger capacity can then be modeled to determine whether it would be financially justified to increase capacity. Strongly interacting products can also be restricted in quantity to learn what the new product mix would look like.

One must be clear about what is desired from an aggregate planning model. It tells what bottlenecks are created by what kinds of product mixes and what constraints and costs seem most to affect profitability. In addition to knowledge of sequence-dependent production times for products, this information can help direct well-conceived schedules that maximize the profitability of resource utilization.

The accuracy of aggregate planning models rests upon the care taken in preparing the resource standards and resource capacities that make up their databases. As was suggested early in this book, development of these databases should be directed by I.E.'s. It also seems reasonable that I.E.'s should direct the running and interpretation of aggregate planning models.

Given a workable aggregate planning model, a number of short-term facility utilization problems can be addressed. Such problems are the scheduling of replacement parts production, rescheduling for new products, rescheduling for new technologies, and the scheduling of rework. Consider what happens when a company that has not been doing so decides to compete with other producers who make replacement parts for its products.

A reasonable assumption might be that, every time x new units of the product are made, a new lot of replacement parts should also be made. The increased rate of production can be studied via the aggregate planning model to determine how the added volume can be accommodated. This same analysis can be performed when a new product is to be added to the production list. However, a new product is a slightly different case.

If a new product is to be added to routine production loading, then its effect must be studied by use of the aggregate planning model. Its time and material standards will have to be largely guessed, but it must be studied so its effect on capacity utilization can be anticipated and prepared for.

Similarly, the introduction of new technological approaches to performing production steps must be studied for their effect on capacity utilization. The new machine or procedure will probably replace an old one that had been a bottleneck, so the entire profile of the plant will be expected to change. New standards, lead times, flow rates, and capacities will have to be guessed prior to startup with the new configuration if operations are to be at all smooth. After startup someone, probably an I.E., will need to study operations to replace the guesses with reliable numbers on which to base aggregate planning and routine scheduling.

Although rework may be implicitly handled by including average rework time in time standards, it may become desirable to make the rework problem more visible by breaking out the time spent on rework as a line item in variance reporting. The data needed to allow for such a move include the fraction of good units that have been reworked on the average along with time standards for the rework functions. Given these data the rework procedures can be calculated as an activity in aggregate planning and in routine scheduling.

Once aggregate planning results have been interpreted into guidelines for production schedules, specific schedules must be developed. For this purpose some version of a simulation model is very useful. The simulation model simply projects the planned schedule over time and determines whether it adheres to capacity limitations. If not, some remedial action must be taken, but the beauty of this application of simulation models is that preventive action, not last-minute reactive action, can be taken with abundant lead time.

SIMULATION

The facilities utilization aspects of production scheduling are complex because capacity conflicts are often complicated, perhaps involving a large number of products. It is therefore difficult to properly anticipate interaction-related bottlenecks without a computational aid. The larger the list of potential products to be scheduled in possible conflict with one another, the larger the computational problem.

Most scheduling simulations are done while holding standards, lead times, and capacities fixed. This type of simulation model is called deterministic. For instance, a scheduler can input a given sequence of jobs to the simulation model, and it will calculate the consequences of that sequence. In particular, a scheduler is usually interested in the total time to complete the schedule, the on-time performance of customer deliveries, and any capacity increases needed to meet due dates.

A simulation model that interacts with a human is sometimes called a decision support system, but there is no need for a specific title to be attached. Modern microcomputers are now equipped with software that can perform deterministic simulations on data input as matrix arrays; these canned computer tools are flexible in their application and may come to replace more complex simulation programs. However, a program that actually performs part of the logic necessary to prepare a schedule that intelligently utilizes facilities would be have to be a complex algorithm.

Stochastic simulations, those that assume probabilistic distributions for times, or costs, or capacities, are not ordinarily used to analyze specific schedules and capacity conflicts. Such models are more commonly used at the design stage to determine whether installed capacity will be adequate. For instance, if a specific machine has been identified as a bottleneck, a revision can be made in its modeled capacity, say doubling it. Then production responses to uncertain demands can be simulated to determine how increased capacity would affect product throughput and profitability on the average.

Recently, it has become increasingly popular to use a more automated version of schedule simulation and capacity analysis. The algorithms involved are modules of the closed-loop P&C systems discussed earlier in this book. They are called a rough-cut capacity planning (RCCP) model and a capacity requirements planning (CRP) model.

ROUGH-CUT CAPACITY PLANNING

The RCCP module of an integrated P&C system interacts with the Master Schedule module to evaluate the capacity utilization required by the master schedule. In the fully automated version the two modules work to revise the master schedule until capacities are not violated. Since a large number of products may be involved, the logic of the master schedule revisions requires a considerable amount of data.

The basic data requirements are gross production resource requirements, gross limitations, product lead times, lot sizes, and priority rules. The module is called '''rough-cut'' because it does not consider every single step in the production process, only the broad outlines. Consider the following small example.

The complete time planning horizon is six weeks. Product A-100 has a master schedule for the period consisting of order quantities with due dates as displayed below in the Table 21.1. Prior production order releases have promised order receipts as indicated in the table. Lead time for a production run is two weeks, and minimum lot size is 10 units of A-100. Starting inventory is 15 units, but no safety stock of the product is required. The calculations summarized in the table show the production schedule (order release and completion schedule) produced by the master scheduling algorithm based on its priority rules.

Table 21.1. Production Scheduling Worksheet for A-100

	Week					
	1	2	3	4	5	6
Order Due Dates	20	10	15	0	20	0
Planned Order Receipts	10	10	0	0	0	0
Inventory 15	5	5	0	0	0	0
Requirements	0	0	10	0	20	0
Production Order Release	10	0	20	0	0	0

The production schedule shows order releases for lots of 10 and 20 units in week 1 and week 3, respectively. The implicit priorities of the scheduling algorithm are:

1. Delay due dates as little as possible.
2. Begin production as late as possible.
3. Minimize inventory.
4. Avoid smaller-than-minimum lot sizes.

Even in scheduling a single product for manufacture there may arise conflicts between these priorities. As soon as a preliminary schedule for production has been derived, it must be checked against resource utilization capacities to determine whether the schedule is feasible. If resource limits are found to be exceeded, some change in the schedule must be made. Usually the choice is of which product to delay, and making that choice requires that product priorities be available.

Product priorities may be fixed, or they may be calculated from data describing current conditions. Some rule that considers production time per unit, proximity of order due date, or cost or profit per unit is commonly used, but an experienced scheduler may develop intuitive priorities as he mentally weighs alternatives.

Let us demonstrate the role of product priorities in fixing the production schedule for two products by use of the RCCP algorithm in the fully automated version that adjusts schedules to conform to resource availability. To perform the calculations the following data will be needed.

Production Order Release Schedule for Two Products

	Week 1	2	3	4	5	6
Product A-100						
Order Release	10	0	20	0	0	0
Product A-200						
Order Release	15	20	0	0	0	0

Resource Requirements for Each Product During Production Schedule

Product	Labor hours/unit Week 1	Labor hours/unit Week 2
A-100	2	2
A-200	2	2

The labor available each week is 40 hours. The labor requirements are called "rough" because they are not real standards but a combination of time requirements from the various steps that make up each week's production activities. Similarly, the hour limit is a rough figure, not being related to any specific work station. The numbers used in the RCCP module of automated planning and control algorithms are simply used to indicate roughly whether a given master schedule can be nominally satisfied within capacities.

In this case the production schedule will require changing because the total labor requirements for the two products are:

Week 1—50 hours (20 for A-100; 30 for A-200)
Week 2—60 hours (20 for A-100; 40 for A-200)
Week 3—80 hours (40 for A-100; 40 for A-200)
Week 4—40 hours (40 for A-100)
Week 5—0 hours
Week 6—0 hours

At this point an automatic algorithm must make use of priorities. Let us assume that Product A-100 has the higher priority. Since A-100 has the higher priority, assume that its schedule will remain as is. Thus the difference between its time requirements and the 40 hours per week is the time left to work on A-200. The respective times left in weeks 1 through 4 are: 20, 20, 0 and 0 hours.

Then product A-200 must have part of its production schedule delayed. It appears that only 10 units of it can be initiated in week 1, and that further production must be delayed until week 5. The earliest production schedule that will provide the number of units required of A-200 is:

Release in week 1—10 units
Release in week 2— 0 units
Release in week 3— 0 units
Release in week 4— 0 units
Release in week 5—20 units
Release in week 6— 0 units
Release in week 7—10 units

Note that the resulting schedule for Product A-200 has a considerable delay to it, running much longer than the planning horizon of six weeks. The order release of 10 units in week 7 also assumes a minimum lot size for the product of 10 units.

If production schedules are set without calculating capacity consequences, this is referred to as infinite-capacity loading. If such a procedure is used, it ordinarily has one of two consequences: (1) resources become very scarce unexpectedly; or (2) someone mentally guesses which orders will be late and attempts to expedite them.

Some amount of expediting is unavoidable, but when it is the rule rather than the exception, the cause is poor scheduling, which is costly and inefficient. Since automated systems for scheduling and RCCP iterations appear to be quite useful, especially when there are many products to schedule, their adoption has become common. However, one must note that no procedures can be successfully computerized that do not work now, and, once adopted, the computer system requires that a voluminous amount of data be prepared and regularly maintained.

To run just the Master Schedule and RCCP modules the following data are required:

Priority rules
Product priorities, static or dynamic
Lot size criteria for each product
Safety stock criteria for each product
Recent inventory updates for each product
Rough-cut resource requirements for each resource for each product
Rough-cut capacities for each resource

Lead times for all products
Current order status reports for each product
Product due dates (firm or forecasted) for all products

This is a large amount of information, but unless it is all available and reasonably accurate, the computerized system will not work well if it works at all. One might note that the Master Schedule/RCCP iterations amount to using a plant loading model in conjunction with logic for revising the production and due date schedules. However, for this computational logic to produce the desired efficiencies with minimal human interference in the decision making, it will require a large effort directed toward maintaining the data properly. An even larger data management problem arises when an MRP/CRP interface is included in a company's panoply of facilities utilization tools.

CAPACITY REQUIREMENTS PLANNING

In a computational sense there is not much difference in the outlines of what Master Schedule/RCCP interfaces do and what MRP/CRP interfaces do. Essentially the same kinds of calculations are performed in each case. The MRP/CRP instance is simply much more detailed.

A CRP algorithm can be visualized as a detailed plant loading model plus the logic that permits it to revise the MRP schedules. The MRP module simply schedules the constituent steps making up the macroscopically viewed products modeled for master schedule calculations. The differences between it and the RCCP program only amount to differences in the level of detail that is modeled.

However, this amount of detail is considerable. If an MRP model is already in place, then lead times, sequences and routings, bills of materials, etc., will be stored in the MRP database. This alone is ordinarily a mountainous array of numbers and other information. Yet, if only the MRP is in place, the procurement and production schedules it produces will be of the infinite-capacity variety.

To install an automatic, schedule-revising CRP algorithm requires a significant increase in the size of the database for every product. Someone, such as an I.E., will have to oversee the gathering of the data intially and then create a system for data maintenance. The benefit a firm gains from installation of a CRP model is the discipline that it forces on operations. This discipline should also translate itself into improved on-time performance, more accurate raw material inventories, better facilities utilization, and reduced WIP inventories.

The data needs are deceptively simple. After all, if an RCCP module is currently in use, there must be capacity and resource requirement data for all products of note. While this argument is correct, the RCCP data is essentially useless for constructing the CRP data. This point is made quite clear when one realizes that product lead times in the RCCP are essentially the sum of lead times for the steps in the product explosion modeled by the MRP/CRP system. Thus, adding CRP capacities to produce RCCP capacities is easy, but deriving CRP data from RCCP data is impossible. It is unlikely that the RCCP model would exist in the absence of the CRP model, however. They are usually installed together.

The CRP algorithm requires that resource standards be developed for every step in the procurement-fabrication-production sequence for every product. Some data needed for the CRP database may be developed during the performance of activity analysis, but the database prepared for activity analysis will probably not be sufficiently detailed to provide a complete base for a CRP system.

Even though CRP installation is a big job, its payoffs can also be large. In particular, consider what will happen if an MRP system is installed without the attendant CRP. The lead times for individual steps will probably be largely guesswork, and as a result, they will probably turn out to be too large, which in turn will make orders for materials to be cut too early, will make raw material inventories too large, and will make WIP inventories too large, also. The collection of scientifically obtained lead times and capacities will not only reduce inventories, they should also avoid capacity conflicts that arise from poorly understood resource limits. However, RCCP and CRP algorithms do have some drawbacks.

CAVEATS

Hundreds of thousands of dollars are usually required just to acquire, adapt, tailor, and install a CRP system. However, there does not have to be an exceptionally large improvement in capacity utilization for the costs to be rapidly recovered. The primary danger is that the company does not now have a planning system that can be adapted to a CRP model or that management will not fully support the continuous monitoring and review such a system requires. It should be the role of I.E.'s to make management aware of these realities.

One further point should be raised. MRP/CRP algorithms promote just-in-time production, but the nature of their approach to conflict resolution tends to ignore the maximization of capacity utilization. Priority rules can remove production from overbooked work stations, but they do not rebook the slack capacity. If maximizing throughput is a significant goal at a manufacturing facility, then a different approach to scheduling may be needed. Vendors currently market several algorithms that include logic to fully load important production steps, usually bottlenecks. An alert I.E. is well advised to familiarize himself with these alternative scheduling and capacity utilization tools.

SUMMARY

Routine scheduling must be planned within capacity availabilities and resource limits. Toward this end several models discussed in earlier chapters can be valuable aids in certain production environments. In repetitive production, aggregate planning models can effectively lay down production schedules. If a considerable amount of human intervention in facilities utilization is planned then simple plant loading models may suffice for resolving capacity conflicts. Complex capacity utilization problems may require elaborate computational algorithms to solve them.

If a plant is planning on adopting the large expensive RCCP and/or CRP algorithms, then I.E.'s are the personnel with the best holistic comprehension of costs and operations to carry off the economic feasibility studies, the systems design, and the implementation strategies upon which algporithm success must be based. I.E.'s should direct the data-gathering effort, since all the data needed is of the kind they well understand. Similarly, I.E.'s should oversee the tailoring, adaptation, installation, and startup of all large-scale P&C models.

22
MARKETING PROCEDURES

Most long-term marketing procedures of interest to strategic planners have already been discussed, but the short-term aspects of these procedures are very important to the profitability of the firm. Although I.E.'s are not usually directly involved in short-term Marketing activities, there are several aspects of the sales function that can benefit from I.E. expertise. To a large degree I.E.'s must educate manufacturing firm managers in all disciplines to the realities of costs and resource utilization. Marketing is no exception, it naturally prefers operations run so as to conform to marketing assumptions, but doing so could impair profitability.

ELASTICITY AND OPPORTUNITY LOSS

Marketing assumptions tend to express themselves as operating rules throughout the firm. To some degree this is inevitable, and insofar as the assumptions can be rationally supported there may be no good reason to disagree with the rules. After all, the sales function generates the revenue. The danger lies in Marketing's promotion of specious ideas that hold costly consequences.

Inventory is a very expensive category of operations that marketing beliefs can affect strongly. Erroneous beliefs create excess stocks because Marketing is wary of stockouts. Salesmen all have stories of customers permanently alienated by product unavailability.

The lost opportunity to make a sale is assumed by almost everyone to have a negative effect on revenues. The translation of this effect into a dollar amount is the act that divides opinion. Salesmen are likely to value the cost of a missed sale at face value; if a $50 pair of shoes is not available when a customer asks for it, a shoe salesman calls it a $50 loss. Other people view it differently.

If the customer returns for the shoes a week later and successfully requests the shoes, the real loss has been $0. Delayed sales are not all lost. Thus, an economist would value the opportunity loss due to not having the asked-for shoes in stock at something less than $50.

However, at a manufacturing firm, if management accepts Marketing's contention as to the higher value of opportunity losses due to stockouts, the tendency is to react by building safety stocks. If on the other hand upper management ignores Marketing's view, safety stocks might be held to a minimum in which case lost sales may actually result. An I.E. studying the situation can see that there should be a rationally tenable compromise between the two extreme approaches to valuing stockouts.

A first step an I.E. would probably take toward resolving the opportunity loss question is the study of historical data to look for patterns between stockouts and demand. If possible, this study would be conducted at several similar distribution points at once. The I.E. would be looking for such behavior as a lagged surge in demand soon after the occurrence of an out-of-stock situation that would indicate the return of undeterred customers. The I.E. would also look for sales that do not rebound after stocks have again become available as evidence of truly lost sales.

As a next step in the study, some method for estimating the sales/inventory tradeoffs should be obtained. Then a rationally based stocking policy can be promoted. It is interesting to note that much the same kind of analysis goes into studying this question as into studying price elasticities.

The questions to be answered are much alike. A price elasticity is a measure of the effect of price on sales volume, while the stockout cost evaluation seeks to measure the effect that inventory levels have on sales volume. Both questions are somewhat interrelated, which fact implies that the two effects on sales should be studied by use of data that avoid mixing the two effects.

Any variation in sales must also be dependent on the level of demand that would exist without stockouts or price changes. Thus the study of opportunity loss, elasticity, and demand are all tied together, and this fact suggests that the design of the database used to study the three should be chosen with all three end uses in mind. Again it appears that an I.E. should lay out an information system, this one for marketing and inventory data collection, so these important questions can be investigated.

DEMAND

Demand forecasting is a critical aspect of manufacturing planning and control. Ideally, the forecasting model should be virtually automatic in operation, able to run from routinely-gathered, timely data provided for in the layout of the forecasting information system. A statistician should have the direct responsibility for maintaining the model, but an I.E. should work with him at the design stage to assure that the model's information system is adequate and to further guarantee that its output will tie in properly with other important information systems in the firm.

It is not at all unusual to find a manufacturing firm whose parts do not communicate with each other. The authors have observed a particular firm whose Market Planning Department and Sales Department worked from completely separate forecasts of demand, each forecast a secret from the other department. It is not at all clear how production and inventory levels are set at that company.

This type of situation is unfortunately not uncommon, which is one reason managers are attracted to closed-loop systems. It is believed that fully integrated information flows will "get everyone on the same page." Toward this end it is a good idea for an I.E. structuring information flows to at least minimally try to tie sales forecasting to other important functions.

Algorithms performing forecast-related functions in many firms are DRP, Inventory Planning and Control, Master Scheduling, and Marketing. If these programs lay out procedures that cross several distinct departments, such as Marketing and Production and Inventory, all distinct entities need to be tied into the information network. The broad outlines of this approach to data management should be provided by corporate strategic planning. Diverse other functions should also be incorporated into the sales/forecasting data network, such as Order Entry/Release.

ORDER ENTRY/RELEASE

This specific function was discussed in some detail earlier in this book. Within a fully integrated P&C system, it might appear as a subset of Production Control, since tracking the progress of production orders is largely motivated by the needs of this function. However, order entry/release is also central to the needs of sales forecasting.

Forecasts must be dynamically refined over time, adjusting to trend, to cyclic change, and to seasonal variation. Past data are the basis for prediction in forecasting models, and models turn out better forecasts when recent, accurate data are provided for them. Especially important are recent sales quantities, which should be generated by the order entry and release function.

Both entry and release are important to planning. The order entry quantities affect the master schedule directly by listing out due dates, and the release data provide the current sales quantities upon which forecast revisions can be based. Numbers that are not current can be virtually useless to planning and forecasting. Currency is also an important requirement for data used in DRP. Distribution requirements planning relies upon detailed order entry/release data, broken down by distribution point.

Even if a company does not opt for the employment of large-scale functional integration, feedback information loops must be designed into the information systems in use. However, this is not the only important aspect of marketing that can use the expertise of the I.E. Another topic with which I.E.'s can assist is that of coordinating Marketing and Production goals within the strategic plan.

SHORT-TERM PRODUCT MIX

At the long-term strategic planning level it may not be difficult to obtain agreement as to the plan's objectives from all parties. But in the heat of battle everyone's viewpoints withdraw into the myopic view of their own disciplines. Then serious conflicts arise. This is when the objective outlook of the I.E. is sorely needed.

Even though Marketing has long understood that it should coordinate activities with Production, pressures of the moment may lead Marketing personnel to strike out on a path of product emphasis or product introduction without alerting Production nor anyone else. Someone, such as an I.E., needs to be on the lookout for this kind of event.

The problems caused by unilateral action have to do with a number of operational issues, including capacity utilization, distribution planning, and profitability. Any shift in product emphasis must be studied for its effect on achieving aggregate goals. Unilateral moves by Marketing may subvert corporate goals and must be intercepted before the damage is done.

Marketing personnel do not ordinarily understand the cost-related concepts of product mix. Thus, they do not realize that arbitrary, sudden changes in the demands placed on production facilities have extensive repercussions throughout the production, inventory, and purchasing chain that can severely limit profits.

The cure for this problem may be to educate Marketing to the fact that significant shifts in product emphasis must be studied for their overall advisability. Like most people not closely associated with production, Marketing personnel commonly hold the view that one product cannot significantly disrupt smooth operations. Only persistent attempts to educate them can hope to change this opinion—a worthy goal for an I.E.

Granted, fairly sudden product moves may be forced upon a company by short-term conditions, but the attempt should be made to meld them into planned production schedules as smoothly as possible. Toward this end someone like an I.E. should try to develop as rational a rapid transition as conditions allow.

Whether the short-term changes are to be made fairly leisurely or very quickly, the ability to plan for production's transition from one product mix to another will be contingent upon the existence of suitably flexible tools for revising plans. Some such tools already mentioned were aggregate planning models, plant loading models, product mix analysis models, activity analysis models, and the entire panoply of P&C models.

In addition to trying to educate Marketing to cost realities, the strategically aware I.E. must be constantly thinking ahead to predict the kinds of emergencies that might arise. Only anticipated problems can be prepared for adequately, and sudden shifts in Marketing emphasis is surely one problem to be ready for.

Within the framework of the strategic plan the I.E. should also be continually seeking ways to cut costs. One prime candidate for savings is to be found in the study of Marketing staffing.

STAFFING

In recent years many companies have found their sales staffs to be either the wrong size or ill-distributed. Dealing with the question of how a department should be staffed is a common activity of I.E.'s and also constitutes the short-term version of the organizational analysis discussed in an earlier chapter.

Sales staff sizes are often more volatile than staffs associated with other functions of the firm. Production and Accounting staffs, for example, may remain quite stable. The difference is that sales staffs must react to entirely different customer needs over time; they must learn to work with new product lines; and they are faced with wide swings in both national and regional demands. Sales management must adapt to these variations while simultaneously trying to minimize costs.

History shows that these needs must be difficult to balance. Many a company has found itself with drastically oversized Marketing/Sales staffs. The problem becomes further compounded by the fact that, once people have been hired, managers find it difficult to fire them; no one looks forward to dismissing people. Corporate fat is hard to pare. I.E. assistance, however, may help reduce overstaffing before it has occurred, through better planning.

During growth periods, when a company sees demand skyrocket and profits climb quickly, it is common for companies to hire ever more Marketing/Sales personnel. Little attention at these times is given to whether they are truly needed; income is rising so fast that the company can afford practically anything. When the market cools is the time when bad management, bad cost control habits, and overstaffing become apparent.

The I.E. alternative to hectic hiring of salesmen and other Marketing employees is to try to estimate work loads, job standards, and the resource capacity that each person performing a job represents. Overstaffing begins by assigning someone to do a job that could have been done by an existing employee—perhaps by allowing someone to work at a slight overload for a short time or by hiring part-time help.

The rational approach to establishing proper office staffing is to study the jobs being performed, decide in each case whether the job needs doing, obtain standards for important tasks, and measure resource capacities against which office loadings can be compared. In the study of sales staffs it may be difficult for someone not fully familiar with marketing/sales procedures to develop the data needed, so it is imperative that someone with authority in this discipline work with a coordinating I.E. to obtain information.

The difficulty of obtaining meaningful measures of the sales force's work load may be the primary reason why sales staff sizes are hard to control. Much of sales work is conducted in privacy between a salesman and his potential customer, and seemingly fruitless contacts may be absolutely necessary to the closing of future deals. An I.E. cannot easily develop objective standards for this kind of activity, so cooperation of marketing/sales management must be obtained if staffing is to be rationalized. The I.E. in this case will probably serve mostly as an educator, showing management what needs to be done and why.

Soliciting cooperation from managers in the various corporate offices that compete with each other for funds and control is not easy. To be successful, an I.E. must be demonstrably objective, not trying to tilt resources toward any particular department, and he must be known for his commitment to the furthering of the corporate strategic plan. He should be known to be interested in cost reduction, and this may be the factor that will most greatly induce cooperation. Every manager that is accountable for a budget is interested in cost control.

A Sales manager may be judged largely by how much revenue his sales force produces per dollar of his budget. If he can be made to understand that extra people on his staff do not imply greater productivity, he may listen and cooperate. As in virtually every area of cost control, I.E.'s need cooperation from the people directly involved, but it is especially necessary in sensitive cost areas such as staffing. Sales personnel must also be cooperative in another problem area if corporate plans are to succeed—the problem area being that of expediting orders.

EXPEDITING

To Production personnel, expediting usually means something has not gone as expected. Since things rarely go exactly as expected, some degree of expediting will always be present in a manufacturing firm. However, it should be held to a minimum. Assigning extra resources of some kind is the basic mode of expediting a late or undersupplied order, and this always costs money, which means that an expedited order actually makes less profit than had been planned.

Marketing/Sales personnel see late orders as a threat to their ability to satisfy customers and promote future sales. In part this is a valid view. However, salesmen rarely look with discrimination at late orders. They do not consider that some orders are more costly to expedite than can be justified. This viewpoint on the part of salesmen is understandable, but it is based in large part on ignorance.

An I.E. might be able to educate sales personnel as to the effect expediting has on profitability. He should also make this point with Production personnel and with upper management. To do so effectively, he will have to make his case from well-developed data. He will have to show the increase in costs per unit of production that each category of expediting incurs.

Many managers do not understand the realities of production processes. For instance, they do not understand that a faster or slower production line does not necessarily produce at a predictable rate when its speed is altered. It is not widely understood that quality control problems multiply when orders are rushed. And it is not well understood that overtime labor costs can turn a small profit into a loss on a given order.

To align everyone's thinking, the I.E. must represent the case for overall corporate profitability. He must demonstrate what costs are associated with expediting, and he should show a few startling examples of how rushing orders through can destroy profits. Perhaps then Production will not be inclined to treat every late job as a major problem and Marketing/Sales will not be so inclined to insert rush orders into an already established schedule.

SUMMARY

The lessons the I.E. learns in dealing with a specific discipline such as sales can be repeated in departments throughout the firm. Every area of specialization will have aspects that are fully understood only by those who work in it all the time. If the I.E., who is trying to enhance the effectiveness of the strategic plan, is to achieve cost and efficiency objectives, he must elicit cooperation from the people in each major area.

The only way that he can expect to gain the ear of naturally suspicious managers and workers is to present a clear record of integrity, objectivity, and success at instigating improvements. Then he may be able to educate the people in the diverse disciplines to the strategic goals he is hoping to help the company attain. In effect, the strategically aware I.E. must show people how things work and why they should be done as he suggests. And he must show that his entire rationale is based upon understandable strategic plans.

PART FOUR

ALLOCATION OF INDUSTRIAL ENGINEERING RESOURCES

Like the physician who can be caught chain smoking, Industrial Engineers are often guilty of not applying the same advice to ourselves that we give to others. How often do we look at our activities to see how productive we are? Are our traditional concerns with labor cost control still appropriate? How do we allocate I.E. resources to an increasingly broad range of productivity related problems? How do we measure our own productivity? Are Industrial Engineering efforts directed toward achieving worthwhile objectives?

Although these questions are difficult to answer, they are too infrequently asked. This book has examined the issues underlying these questions and other issues concerning the productivity of Industrial Engineering. Suggestions for improving the productivity of Industrial Engineering through a new charter will now be made.

The charter and priorities of Industrial Engineering must be revised completely. There are too few industrial engineers to permit them to be underutilized on unproductive or minimally productive jobs. Industrial Engineering is a scarce resource and must be managed as any scarce resource must be to obtain the most significant output per unit of input.

The priority action item that Industrial Engineering must adopt as a profession is to have each of its members commit himself and accept the responsibility for total productivity improvement. Complacency and satisfaction with minor or limited productivity objectives must end. The need to change priorities is not altogether due to the failing of Industrial Engineering. Management has also been responsible for allowing a scarce human resource to slide into less than fully productive use, but the momentum toward change will undoubtedly have to come from the Industrial Engineering profession itself.

Industrial engineers must become involved in those functions of their companies that define their competitive strategies. This involvement can be achieved in various ways, some of which have been suggested throughout this book. But first, industrial engineers must come to view their own function as a valuable, scarce strategic resource.

23
A NEW CHARTER FOR INDUSTRIAL ENGINEERS

Industrial Engineering is usually estimated to have had its beginnings with the teachings of Frederick Taylor, who believed management, especially management of labor resources, could be put on a scientific basis. He further believed that a scientific approach to work would increase productivity and drive down costs of production, benefiting capitalist and worker alike.

I.E.'s still follow Taylor's teachings, still pursue productivity improvements and cost control. Unfortunately, it seems the profession may not have kept pace with changes in the manufacturing environment. Taylor's science was primarily addressed toward solving the problems of labor-intensive operations. Today's I.E.'s are faced by a range of issues such as those discussed throughout this book.

The authors believe that to a large degree the I.E. profession is still addressing the problems of Taylor's day and that the thrust of I.E. efforts has fallen behind the times. For industrial engineers to exert themselves in the ways that will achieve the most good, we believe that a new charter, a more modern definition of the I.E. function, must be adopted. I.E.'s must partake in the strategic planning and implementation activities of manufacturing firms throughout their operations. The availability of I.E.'s is too small to permit less than their full utilization.

INDUSTRIAL ENGINEERING AS A SCARCE RESOURCE

When one examines a production flow chart, one looks for the scarcities that may restrict production. A scarce resource can demand a high marginal value, i.e., the value of an additional unit of that resource. The critical resources will always have the highest marginal values. Industrial Engineering has become a scarce resource for U.S. industry.

The proof of this fact is easy to demonstrate. In the face of an estimated 8000 job openings for industrial engineers each year slightly more than 3000 graduate industrial engineers are being produced per year. Despite this shortfall, engineering schools all over the U.S. are restricting enrollment. Disregarding the current economic conditions, this means that the 5000 Industrial Engineering jobs per year that go begging now will only become more numerous.

Industry is hoping to improve productivity and profitability, yet productivity personnel have become a scarce resource. What jobs should I.E.'s be working on? Only jobs with potential returns comparable to the marginal value of their time; I.E.'s should work only on the jobs that address the most important strategies of the company. How are industrial engineers actually being used? Much differently.

THE MISUSE OF THE I.E. RESOURCE

What activities should Industrial Engineering be directed toward? Characteristically, industrial engineers have been involved with direct labor, primarily in manufacturing. However, as Table 23.1 shows, the importance of this sector of the work force has been declining. At the same time the costs of energy, materials, and capital have skyrocketed, yet the efforts of industrial engineers are directed primarily at the direct labor segment of the work force.

Table 23.1 Distribution of Employment

Occupational Group	1968 Percent of Work Force	1980 Percent of Work Force
Total White Collar Workers	46.7	50.8
Professional/Technical	13.6	16.3
Managers/Officials	10.0	10.0
Clerical	16.9	18.2
Sales	6.0	6.0
Total Blue Collar Workers	36.3	32.7
Craftsmen/Foremen	13.1	12.8
Operators	18.4	16.2
Laborers	4.7	3.7
Service Workers	12.4	13.8
Farm Workers	4.6	2.7

Source: U.S. Department of Labor Bulletin 1637

Many industrial engineers find themselves maintaining direct labor standards or doing other low-productivity clerical work. Industrial Engineering frequently falls into the "systems maintenance trap." After creative I.E.'s have developed productivity-enhancing systems they are often rewarded by being assigned the clerical task of maintaining the systems with the result that a highly productive resource is directed toward a much less productive task.

HOW CHANGE MUST COME ABOUT

Change cannot come abruptly. There is too much inertia of habit, and the change cannot be expected to come from upper management; habitual attitudes toward I.E.'s will not shift spontaneously. The change will have to be initiated, pursued, and implemented by industrial engineers themselves.

The objective is clear: I.E.'s must become fully involved in the most important activities of the firm. They must take control of and responsibility for the implementation of the dominant aspects of competitive strategy. Failure to do so may result in a permanent misuse of the scarce Industrial Engineering resource and a simultaneous loss of competitive edge throughout American industry.

To take this role, I.E.'s must show initiative. They must dedicate themselves to control of costs and improvement of productivity. They must educate themselves to corporate strategy. Then they must propose projects for themselves that will make solid contributions to profitability. They must further take responsibility for the projects they propose.

There are two usual responses to this suggestion. Working industrial engineers say, "If I make proposals like that, I'll just get shouted down or ignored."

Managers say, "I would welcome such initiative from my industrial engineers. In fact, I have been trying to encourage it."

It is to be seriously doubted that intelligent initiatives will be consistently rejected. However, any company that is unwilling to accept such intitiative does not deserve to retain bright industrial engineers, who should apply their skills where they will be appreciated.

The problem facing the aggressive industrial engineer who has perhaps been immersed in maintaining standards, is how to identify the most productive projects. This is not likely to be a critical problem at first, since a good I.E. can spot many activities in which he can make a marked improvement, but eventually his goal should be to have an input at the highest corporate level. To do so he must become familiar with some of the common approaches to corporate strategy and apply them to the development of his own strategy. Toward this end the prior chapters have treated that very topic.

STRATEGIES FOR INDUSTRIAL ENGINEERING

Much as in the corporate application of strategic planning tools, industrial engineers can use strategic planning tools for determining how to allocate their own scarce resources. Figure 23.1 shows how an allocation of productivity resources might follow a strategy consistent with that of the company as a whole.

In Figure 23.1, the circle sizes represent the relative investment of productivity resources allocated to the different products. The mature products, the cash cows (quadrant A), should be allocated limited productivity resources since the company already maintains a dominant cost position in this product category. Industrial Engineering efforts in this case might be devoted primarily to the maintenance of the process to assure that the cost advantages are not lost.

Products in the star category (quadrant D) are allocated the bulk of the productivity resources. These are the products upon which the business rests its future. Industrial Engineering needs to take a major role in cost reduction for these products. The object is to ensure the company the lowest possible cost position as the product matures and the growth rate declines. I.E. efforts need to be devoted to such activities as developing automated technologies, making methods improvements and seeking other major process improvements.

Significant productivity resources should also be devoted to a few of the more promising question mark products (quadrant C). The objective is to achieve reductions in the most promising products' costs. Industrial engineering activ-

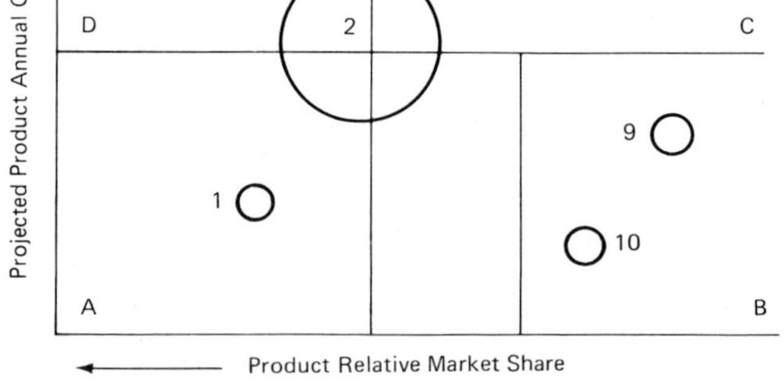

Figure 23.1. Allocation Strategy for Productivity Improvement Efforts.

ities should be devoted to initial process development in anticipation of future efforts at full scale operation.

Products in the dog category (quadrant B) should be allocated few productivity improvement resources. Since the company should be phasing out these businesses, productivity resources might at most be devoted to minor maintenance, investment recovery, and the steps required to phase out the product.

The conceptual strategy for allocating I.E. resources that evolves from the discussion above can be outlined as follows:

1. The potential for cost reduction among different products and processing steps can be identified from cost studies of those products and steps.
2. The allocation of I.E. resources to different activities should be done in a way consistent with the organization's strategic plan and in a way that maximizes the utilization of the scarce I.E. resource.
3. Measurement of the productivity of Industrial Engineering resources and activities should be based on how the improvement of the company's cost position, especially as it compares with that of its competition.

It is important to note that this is a conceptual approach to allocating industrial engineering resources. It outlines in general how Industrial Engineering's productivity improvement efforts can come to be more in tune with the overall direction of the business. There are, however, a number of specific productivity issues that need to be addressed.

MEASURES OF PRODUCTIVITY FOR INDUSTRIAL ENGINEERING

It is ironic that Industrial Engineering, the productivity profession, has done little to measure its own productivity. Occasionally one hears statements such as: "An industrial engineer should save the company X times his salary in any given year." Investigation reveals little evidence that this criterion is being applied.

The problem that a singular measure such as this one presents is that it fails to consider the total activities of an Industrial Engineering department. What is needed are several measures that give a better picture of what Industrial Engineering is actually accomplishing. Three types of measures are needed:

1. Measures that deal with the total performance of the Industrial Engineering Department.
2. Measures that relate to specific aspects of the Industrial Engineering Department's performance, but which are difficult to translate into cost reductions.

3. Measures which relate to specific aspects of an Industrial Engineering Department's performance that can be interpreted as cost reductions.

The first type of measure needs to encompass the total industrial engineering program. While standard productivity measures (e.g., man-hour/unit) could be used, they have a major limitation in that they tend to be restricted to only the labor component of productivity. A better measure might be the value-added cost per unit produced. The cost per unit in this case should include both direct costs and indirect costs. The industrial engineering program must begin to seize responsibility for the total costs of the organization, and use of this measure would further that objective.

The second series of measures describes activities such as writing job descriptions, managing training programs, designing information systems, updating standards, coordinating operating systems, and developing incentive pay systems. For each of these items it is difficult to measure the productivity of industrial engineering efforts strictly on a cost reduction basis. Performance in these job categories should be measured according to specific goals. For example, a goal for updating standards might be: revise 140 standards using no more than an average of four hours per standard. The goal in this case should relate to both the quantity of the activity and the resources budgeted for it.

A third set of measures deals with specific cost reduction for every dollar invested in industrial engineering resources. The measurement of productivity in this area should address both the cost reductions accomplished and those that could be realized if proposed steps were implemented.

All three sets of measures are important in the evaluation of the productivity of the industrial engineering function. They form an important source of documentation for the industrial engineering effort that can be used in expanding the industrial engineering charter.

THE CHARTER OF INDUSTRIAL ENGINEERING

Frederick Taylor's original proposal of scientific management was broad in its concept. Industrial Engineering evolved from this beginning and began by concentrating on the direct labor component of production as the key to productivity improvement in a labor-intensive era. This was a proper emphasis at first, since direct labor was a major component of total product costs. The charter of Industrial Engineering (either formal or informal) gradually became confined to direct labor activities.

As Industrial Engineering became successful in reducing direct labor costs, the potential for cost reduction in direct labor activities began to diminish relative to that in other components of the productivity picture. Industrial engineering has in effect reduced its potential by becoming successful. When Industrial Engineering departments began looking to expand into other activities, they were told that their efforts should continue to be directed toward labor-related activities. Gradually, Industrial Engineering evolved from a problem-solving profession into a clerical profession as more and more Industrial Engineering work was devoted to the mere updating or maintenance of systems that had been developed years before.

The need for a new charter is Industrial Engineering's top priority, and this fact is beginning to be recognized at many Industrial Engineering departments. However, a paradoxical situation exists in many of these departments; they are frequently heard to complain that they are constantly rebuffed in their efforts to broaden their efforts. At the same time, higher level management in the same organizations is heard to complain about the unproductive efforts of their Industrial Engineering departments. In one case, a division manager of a large electronics manufacturer told the authors that one of his first activities when he became division manager was to get rid of the Industrial Engineering Department in his division. The question that must be resolved is how can Industrial Engineering present its case for a broadened charter more effectively?

One step toward gaining acceptance of a broadened charter is the commitment on the part of I.E.'s to the realization of productivity improvement cost reduction goals. Strategies for accomplishing these goals need to be developed as discussed throughout this book, and measurable objectives need to be set. Most importantly, Industrial Engineering departments need to accept the responsibility for seeing that improved operations and cost reduction procedures are implemented. When productivity, efficiency, and cost goals are not realized, the Industrial Engineering Department naturally blames those who failed at implementation. But if industrial engineers were more willing to set cost reduction goals and then accept the responsibility for implementation, chances for a broadened industrial engineering charter would be enhanced.

A second step needed to broaden industrial engineering efforts is the development of a project evaluation system for all proposed activities. Industrial Engineering must begin to think of each of its activities in terms of potential payback. Activities devoted to the maintenance of systems must be evaluated in the same way new projects are evaluated. The allocation of I.E. resources must be based more on potential payback than on tradition. The development of a measurement system capable of comparing the potential of different activities would enhance the acceptance of a broadened charter as Industrial Engineering undertakes projects with higher potential.

A third step toward broadening the industrial engineering charter is that of simply becoming more aggressive in seeking out new projects. Industrial Engineering departments in many organizations we have surveyed are perceived as being content to simply maintain systems developed in the past. But it also appears that management is more than willing to grant additional responsibility to industrial engineers if they will make the effort to seek this responsibility.

TRANSFORMING THE INDUSTRIAL ENGINEERING FUNCTION

As Industrial Engineering begins to broaden its charter, there will be changes necessary within the company that in turn will affect the discipline's function. First and perhaps most important, the industrial engineering function must participate in the strategic planning of the business. In many cases today, Industrial Engineering is only involved in the operational side of the business. Because of this, it has not had as great an impact on the total productivity of organizations as is possible. In many cases, strategic level decisions are made that are inconsistent with operational realities. Industrial Engineering involvement at the strategic planning level would help avoid such conflicts by providing solid bottom-up analysis.

To make Industrial Engineering heard at the strategic level will require aggressive selling; it must become recognized for its broader potential. It becomes increasingly important that industrial engineers take on more significant and far-reaching projects. The image of Industrial Engineering as a clerical function must be dispelled. Some potential projects may have to be rejected if they do not satisfy the productivity criteria outlined earlier. While this may seem to be an unrealistic position to take, it is precisely the kind of position an industrial engineer takes after evaluating the feasibility of purchasing a proposed new piece of equipment and finding it wanting.

There is a significant amount of industrial engineering staff resistence to the broadening of the Industrial Engineering function. The clerical function that many industrial engineers now practice is a comfortable one. Certainly clerical work is less risky than is the project-oriented, cost reduction and efficiency improvement program proposed. In many industrial engineering organizations we have surveyed, however, the level of motivation is quite low simply because of the clerical perception of the work being done. While the transformation to a project-oriented program will be uncomfortable for many industrial engineers, the ultimate result of the transformation should be a more motivated staff possessed of a greater feeling of responsibility.

The transformation process will certainly be difficult and traumatic for industrial engineering staffs and challenging for those in leadership positions. The productivity of Industrial Engineering, however, must be improved if America is to better utilize this scarce resource.

As the profession which promotes itself as the productivity profession, we should hold our own high productivity as our priority. We must begin to set an example for others if we are to be successful in our overall productivity objectives.

SELECTED BIBLIOGRAPHY

BOOKS

1. Abell, Derek F. and John S. Hammond, *Strategic Market Planning,* Prentice-Hall, Inc., Englewood Cliffs, NJ, 1979.
2. Ansoff, H. Igor, *Strategic Management,* Halsted Press, New York, 1979.
3. Baker, K.R., *Introduction to Sequencing and Scheduling,* John Wiley & Sons, New York, 1974.
4. Buffa, E.S., *Modern Production/Operations Management,* John Wiley & Sons, New York, 1983.
5. Bussey, L.E., *The Economic Analysis of Industrial Projects,* Prentice-Hall, Inc., Englewood Cliffs, NJ, 1982.
6. Byrd, J., Jr. and L.T. Moore, *Decision Models for Management,* McGraw-Hill Book Co., New York, 1982.
7. *Corporate Planning: Techniques and Applications* (eds. R.J. Allio and M.W. Pennington), AMACOM, New York, 1979.
8. *Corporate Strategy* (ed. Thomas H. Naylor), North-Holland Publishing Company, New York, 1980.
9. Deming, W. Edwards, *Quality, Productivity, and Competitive Position,* Center for Advanced Engineering Study, Cambridge, Mass., 1982.
10. Godiwalla, Yezdi H., *Strategic Management,* Praeger Publishers, New York, 1983.
11. Haas, Raymond M., Richard I. Hartman, John H. James, and Robert R. Milroy, *Long-Range Planning for Small Businesses,* Indiana University Press, 1964.
12. Hitomi, K., *Manufacturing Systems Engineering: A Unified Approach to Manufacturing Technology and Production Management,* Taylor & Francis, Ltd., London, 1979.
13. *Implementing Operations Research/Management Science* (eds. Randall L. Schultz and Dennis P. Slevin), American Elsevier Publishing Company, Inc., New York, 1975.
14. Jay, Anthony, *Management and Machiavelli,* Bantam Books, New York, 1978.
15. Juran, J.M. and Frank M. Gryna, Jr., *Quality Planning and Analysis* (2nd ed.), McGraw-Hill, New York, 1980.
16. Kollat, David T., Rodger D. Blackwell, and James F. Robeson, *Strategic Marketing,* Holt, Rinehart, and Winston, Inc., New York, 1972.
17. Koren, Yoram, *Computer Control of Manufacturing Systems,* McGraw-Hill Book Co., New York, 1983.
18. Levitt, Theodore, *The Marketing Mode,* McGraw-Hill Book Co., New York, 1969.
19. Lovange, Peter and Richard F. Vancil, *Strategic Planning Systems,* Prentice-Hall, Inc., Englewood Cliffs, NJ, 1977.

20. Machiavelli, Nicolo, *The Prince,* Airmont Publishing Co., Inc., New York, 1978.
21. Migliore, R. Henry, *An MBO Approach to Long-Range Planning,* Prentice-Hall, Inc., Englewood Cliffs, NJ, 1983.
22. Ouchi, W.G., *Theory Z: How American Business Can Meet the Japanese Challenge,* Avon Books, New York, 1981.
23. Parkinson, C. Northcote, *Parkinson's Law,* Ballantine Books, New York, 1976.
24. *Perspectives on Experience,* The Boston Consulting Group, Inc., Boston, 1972.
25. Peter, Laurence J. and R. Hull, *The Peter Principle,* Bantam Books, New York, 1970.
26. Plossl, G.W., *Manufacturing Control: The Last Frontier for Profits,* Reston Publishing Co., Inc., Reston, VA, 1973.
27. Porter, M.E., *Competitive Strategy: Techniques for Analyzing Industries and Competitors,* The Free Press, New York, 1980.
28. *Product Line Strategies* (ed. Earl L. Bailey), Conference Board Report No. 816, The Conference Board, Inc., New York, 1982.
29. Purcell, W.R., Jr., *Understanding a Company's Finances: A Graphic Approach,* Houghton Mifflin Co., Boston, 1981.
30. Putnam, A.O., *Management Information Systems: Planning, Development, Managing,* Herman Publishing, Inc., Boston, 1977.
31. Root, Franklin R., *Strategic Planning for Export Marketing,* International Textbook Company, Scranton, PA, 1966.
32. Skinner, Wickham, *Manufacturing in Corporate Strategy,* John Wiley & Sons, New York, 1978.
33. Sloma, Richard S., *No-Nonsense Management: A General Manager's Primer,* Macmillan Publishing Co., Inc., New York, 1977.
34. Steiner, G.A., *Strategic Planning: What Every Manager Must Know,* The Free Press, New York, 1979.
35. Sutton, C.J., *Economics and Corporate Strategy,* Cambridge University Press, New York, 1980.
36. Taylor, Frederick W., *The Principles of Scientific Management,* W.W. Norton & Co., Inc., New York, 1967.
37. Woolsey, R.E.D. and H.S. Swanson, *Operations Research for Immediate Application: A Quick & Dirty Manual,* Harper & Row, Publishers, New York, 1975.

NEWSPAPER, MAGAZINE, AND JOURNAL ARTICLES

38. Bodde, David L., "Riding the Experience Curve," *Technology Review,* March-April, 1976, pp. 53-59.
39. Bourke, John T., "Performance Bonus Plans: Boon for Managers and Stockholders," *Management Review* 64, November, 1973, pp. 13-18.
40. Black, J.T., "Cellular Manufacturing Systems Reduce Setup Time, Make Small Lot Production Economical," *Industrial Engineering,* Vol. 15, No. 11, November, 1983, pp. 36-48.
41. Byrd, Jack, Jr. and L. Ted Moore, "I.E.'s Skills Equip Them to Effectively Build and Implement Plans for Corporate Strategic Activity," *Industrial Engineering,* Vol. 14, No. 10, October, 1982, pp. 34-39.
42. Byrd, Jack, Jr. and L. Ted Moore, "The Application of a Product Mix Linear Programming Model in Corporate Policy Making," *Management Science,* Vol. 24, No. 13, September, 1978, pp. 1342-1350.
43. Conley, W. D. and F. W. Miller, "MBO, Pay, and Productivity," *Personnel* 50, January-February, 1973, pp. 21-25.

44. Cowdrick, Robert W., Jr. and Michael Spencer, ''Business Reporting Software Systems,'' *Production and Inventory Management Review,* May, 1985, pp. 62-65.
45. Delombre, J. and B. Bruselius, ''Importance of Relative Market Share in Strategic Planning,'' *Long Range Planning,* August, 1977, Vol. 10, No. 4, pp. 2-7.
46. Ein-Dor, Phillip and Eli Segev, ''Strategic Planning for Management Information Systems,'' *Management Science,* Vol. 24, No. 15, November, 1978, pp. 1631-1641.
47. Fierman, Jaclyn, ''A Safe Way to Test Financial Fantasies,'' *Fortune,* April 29, 1985, p. 58.
48. French, Robert L., ''Management Must Face Issues and Realities in Implementing a CIMS,'' *Industrial Engineering,* Vol. 16, No. 8, August, 1984, pp. 70-77.
49. Hax, Arnoldo C. and Nicolas S. Majluf, ''The Corporate Planning Process,'' *Interfaces,* Vol. 14, No. 1, January-February, 1984, pp. 47- 60.
50. Javetski, Bill, ''Dome Finally Gets a Handle on Its Debt,'' *Business Week,* February 18, 1985, pp. 130-132.
51. Latham, Gary P. and Gary A. Yukl, ''A Review of Research on the Application of Goal Setting in Organizations,'' *Academy of Management Journal* 18, December, 1975, pp. 824-845.
52. Levitt, Theodore, ''Exploit the Product Life Cycle,'' *Harvard Business Review,* November-December, 1965, pp. 81-94.
53. Merrill, H.M. and F.C. Schweppe, ''Strategic Planning for Electric Utilities: Problems and Analytic Methods,'' *Interfaces,* Vol. 14, No. 1, January-February, 1984, pp. 72-83.
54. Miller, Danny and Peter H. Friesen, ''Archetypes of Strategy Formulation,'' *Management Science,* Vol. 24, No. 9, May, 1978, pp. 921- 933.
55. Mintzberg, Henry and Melvin F. Shakun, ''Introduction to a Grouping on Strategy Formation,'' *Management Science,* Vol. 24, No. 9, May, 1978, p. 920.
56. Mintzberg, Henry, ''Patterns in Strategy Formation,'' *Management Science,* Vol. 24, No. 9, May, 1978, pp. 934-948.
57. Mintzberg, Henry, ''The Manager's Job: Folklore and Fact,'' *Harvard Business Review,* July-August, 1975, pp. 49-61.
58. Murray, Edwin A., Jr., ''Strategic Choice as a Negotiated Outcome,'' *Management Science,* Vol. 24, No. 9, May, 1978, pp. 960-972.
59. Muther, Richard and Edward J. Phillips, ''How Facilities Planners View the Planning Process,'' *Industrial Engineering,* Vol. 15, No. 3, March, 1983, pp. 44-48.
60. Myers, Stewart C., ''Finance Theory and Financial Strategy,'' *Interfaces,* Vol. 14, No. 1, January-February, 1984, pp. 126-137.
61. ''New Breed of Strategic Planner,'' *Business Week,* September 17, 1984, pp. 62-68.
62. Patton, Arch, ''Industry's Misguided Shift to Staff Jobs,'' *Business Week,* April 5, 1982, pp. 12-15.
63. Potts, Mark, ''New Planning System Aims to Boost Speed, Flexibility,'' *Washington Post,* September 30, 1984, Section K, p1ff.
64. Powell, Wialiam J., Jr., ''Exxon: A Drive to Disaster in the Office,'' *Business Week,* June 3, 1985, pp. 94-95.
65. Rockhold, Alan G., ''White Collar Productivity: Top Management Doesn't Know What It Wants,'' *Infosystems,* April, 1982, pp. 58-64.
66. Schendel, Dan and G. Richard Patton, ''A Simultaneous Equation Model of Corporate Strategy,'' *Management Science,* Vol. 24, No. 15, November, 1978, pp. 1611-1621.
67. Schroer, Bernard J., J.T. Black, and Shou Xiang Zhang, ''A Just- In-Time Small Batch Manufacturing System,'' *Industrial Engineering,* Vol. 16, No. 6, June, 1984, pp. 54-65.
68. ''Scoreboard Special,'' *Business Week,* March 22, 1985, pp. 164-194.
69. Skinner, Wickham, ''The Focused Factory,'' Harvard Business Review, May-June, 1974, pp. 113-121.

70. Skinner, Wickham, "Operations Technology: Blind Spot in Strategic Management," *Interfaces,* Vol. 14, No. 1, January-February, 1984, pp. 116-124.
71. Steinbreder, H. John, "Ma Blue: IBM's Move Into Communications," *Fortune,* October 15, 1984, pp. 52-58.
72. Sweeney, Dennis J., E.P. Winkofsky, Probir Roy, and Norman R. Baker, "Composition vs. Decomposition; Two Approaches to Modeling Organizational Decision Processes," *Management Science,* Vol. 24, No. 14, October, 1978, pp. 1483-1490.
73. "Texas Instruments Shows U.S. Business How to Survive in the 1980's," *Business Week,* September, 18, 1978, pp. 66-70.
74. Wheelwright, Steven C., "Strategy, Management, and Strategic Planning Approaches," *Interfaces,* Vol. 14, No. 1, January-February, 1984, pp. 19-33.
75. Zannetos, Zenon S., "Strategies for Productivity," *Interfaces,* Vol. 14, No. 1, January-February, 1984, pp. 96-102.

INDEX

INDEX

A,B,C classification
  of inventories, 209, 214, 217
  of customers, 227
Accountability
  for R&D, 84
  in decision structure analysis, 148
  in organizational structure, 17, 92, 146
  lack of as a corporate detriment, 87, 88
Accounting, 16
  as a basis for activity identification, 56
  as CIMS module, 199
  managerial, 57, 313
  role in budget monitoring, 102
Acquisition, 9
  as a paper activity, 79, 107
  as vertical integration, 14
  characterization of, 67
  consequences of, 10, 75, 78, 80
  debt resulting from, 78
  discrete, 76
  I.E.'s role in, 10, 78
  marginal, 76
  of distribution warehouses, 122
  resulting management structure from, 77
  reasons for, 9, 135
  unfriendly, 77, 78
Activities identification, 55, 56
Activity analysis, 55, 235
  applied to distribution planning, 122
  in budgeting, 98, 102
  in long-term planning, 63
  related to product mix analysis, 172
Activity mix
  as explained by economic reasoning, 58
  derived from corporate objectives, 54
  for two plants, 78
  optimal, 54
  vs. product mix,63, 172
Allocation of costs, 56
  binding constraints in, 59, 61
  full, 57
  in supply sourcing, 119
  labor-hour basis for, 57
Amdahl, 11
Assembly lines, 113
AT&T, 19
Automobile firms
  Japanese, 75, 100, 132, 156
  U.S., 13, 100, 114, 132, 156
Availability
  as a function of redundant design, 186
  as a plant performance criterion, 186
  calculation for parallel design, 187

Backward explosion of schedules, 177, 198
Bar coding, 207
Boston Consulting Group, xi
Bottlenecks
  as limitations on product mix, 64
  identification from planning models, 236
  in expansion analysis, 70
  in facilities loading, 235
  in product development, 165
  motivating methods choices, 25
Bottom-up analysis, 35
  as a basis for focus, 54
  as a basis for plan implementation, 35
  based on strategic agenda, 136

Bottom-up analysis (*Continued*)
data for, 34
in budget planning, 98
in capacity analysis, 44
in distribution planning, 122, 125
in expansion/acquisition studies, 68
in facilities decisions, 43
in market research, 86
in PM scheduling, 189
in product mix studies, 182
in redundant design, 189
in supply sourcing, 117
Breakeven point, 51, 135
British Steel, 46
Budgeting, 97
as a criterion in organizational research, 92
as an indicator of poor planning, 37
as an operational necessity, 12
as based on activity mix, 99
as based on revenue predictions, 98
automatic increases in, 98
based on standards, 100
cash shortages in, 98
conferences on, 101
contingency resources in, 99
cost basis for, 13
for research, 87
in finance strategy, 12, 97
in product development, 164, 165, 166, 167, 168
irrational, 98
lack of, 98
of market development, 94
of staff functions, 100
strategic principles of, 97
variance reporting in, 102
Buffer stocks
due to facilities utilization, 235
in distribution decisions, 125
in inventory policy, 118
in supply strategy, 112
in U.S. factories, 185
Business Week, xi, 11

Caesar, Julius, 111
C.E.O.
delegation by, 149
R&D education of, 86
role in budget debates, 101
role in corporate philosophy, 34
role in strategic planning, 143
Capacity analysis, 44, 178
applied to CRP models, 243
applied to RCCP models, 243
as a basis for goal setting, 45
I.E.'s role in, 45
in facilities utilization, 236
in marketing strategy, 15
with aggregate planning models, 237
Capacity requirements planning (CRP), 242
cost of installation, 243
I.E.'s role in, 244
interaction with MRP, 242
module of CIMS, 198
Capacity utilization, 43
case study of, 48
efficiency improvements from, 47
I.E.'s role in, 47
managerial misunderstanding of, 47
prediction of, 62
Capital costs, 65, 103
effect on marketing strategy, 135
in feasibility study, 70
Capital intensive operations, 103
Capital productivity, 64, 103
I.E.'s role in, 103
measures of, 65
Cash cow, 139, 158
bank checks example, 191
concept applied to replacement analysis, 190
reductio ad absurdum, 140
Cash flow
definition, 191
expected, 190
in replacement/retirement analysis, 191
in ROI calculation, 65
projected for investment analyses, 50
retained, 103
CIMS, see computer integrated manufacturing systems
Coleco products, 131
Adam computers, 23, 164
Cabbage Patch Kids, 23
Compaq, 11
Computer integrated manufacturing systems (CIMS), 30, 193
alternatives to, 32
as just-in-time systems, 197

automation of, 199
I.E.'s role in, 31, 200
in scheduling, 225
modules of, 195
requirements of, 230
scheduling in, 230
Computer Science
as a poor division to supervise CIMS, 201
as a staff function, 16
in scheduling system selection, 231
Computerization
adaptation and debugging of, 184
affecting corporate culture, 8
and management, 200
as closed-loop systems, 230
diverse disciplines in, 184
DSS approach to, 32
I.E. role in, 31, 184
in inventory control, 26, 218
limitations of, 30, 149, 159
MRP systems, 8
of deterministic simulation, 238
of distribution analysis, 127
of inventory control, 214
of order entry/release, 146
of production planning and control, 30, 195
of scheduling, 224, 229
strategies of, 31
successes of, 184
training aspects of, 31, 184
Computers
as attempted replacement for people, 146
changes in, 132
faddish popularity of, 149
in activity analysis, 62
in aggregate planning, 236
in decision support systems, 229
in plant loading analysis, 176
operational use of, x
Conflict resolution between corporate divisions, 225, 247
Conglomerates, 135
Consensus management, 142
Consultants, x, 190
Consumption of capital, 103
Continuous casters, 4, 69
Contribution margin (P/V), 57
Control charts, 90
Coordination of functions, 164, 165, 166, 167
Corporate culture
as a cause of poor focus, 8
changes in, 8
ingrained attitudes in, 8
philosophy in, 136
research aspects of, 88
role in acquisitions, 77
Corporate obesity, 144, 150
Corporate philosophy
I.E.'s detection of, 136
IBM's, 136
implicit vs. explicit, 136
Corporate strategic planning
as a guide to company structure, 142
as a selection of futures, 85
as affected by organization, 141
as affected by technological change, 51, 237
as dictating product mix, 172
as the direction for product development, 163
at General Electric, xi
budgeting in, 92, 97, 98
C.E.O.'s responsibility for, 143
demand fluctuations effect on, 134
derived from focus of direction, 54
distribution in, 14, 121
effect on methods of production, 183
facilities decisions in, 3, 43
failures of, xiii, 33, 35
finance component of, 4, 12, 99, 107
growth assumptions in, 67
I.E. involvement in, xi, 37, 39, 257
implementation of, xi, 35, 38
in integrated P&C systems, 194
in inventory policies, 213, 221
level of, xi
line experience in, xi, xiv, xv
long-term aspects of, 1
market development in, 93
objectives in, 34, 127
outside financing in, 109
profitability as a motivation of, xiv, 101
provision for short-term models in, 182
R&D in, 11
realistic basis for, 33
research in, 84, 85
short-term aspects of, 1, 3, 19, 153
Strategic Planning Manager in, 143
subclassifications within, 1
top-down aspects of, 34

Cost analysis
based on capacity analysis, 55
based on standards, 55
for transportation networks, 129
I.E.'s role in, 55, 102, 119
in market research, 86
Cost center decisions, 18, 79, 80
Cost control, 138
as a motivation for staff reduction, 147
as goal of I.E.'s, 157, 255
at plant level, 158
planning of, 139
Cost cutting strategies, 4
in mature markets, 132
in product positioning, 138
staff reduction in, 62
Cost reduction, 87, 89
coordination of, 167
factors in, 158, 159
I.E.'s role in, 139
in assessment of I.E. effectiveness, 259
planned, 139
through product rollover, 139, 159
Costs
allocation of, 55, 56, 158
as a basis for focus of direction, 55
as a dominant element of strategic planning, xii
as a reason for vertical integration, 114
as a supply source criterion, 112, 119, 120
based on machine hours, 191
estimation, 102, 161
in distribution planning, 121, 181
in product positioning, 157
in retirement/replacement analysis, 191
in staff sizing, 148
information systems in, 139
monitoring of, 158, 159, 161
of automation, 184
of downtime, 185
of expediting orders, 251
of labor, 138
of lost orders, 245
setup, 216
sunk, 190
tracking of, 56
tradeoffs, 30, 186
variable, 172
Darwinian selection, 107, 115, 158
Databases
for CRP model, 242
for P&C system, 195
for aggregate planning models, 236
for preventive maintenance planning, 188
for short-term computational tools, 21
DCFROI (Discounted Cash Flow ROI) in investment decisions, 50, 66, 75
Debt, 97, 102, 104, 109
Decision making levels, 141, 142, 149
Decision structure analysis, 148, 149
Decision support systems (DSS), 204, 238
in distribution planning, 203
in resolving P&C conflicts, 198
in scheduling, 229
intuition in, 204
Demand
as input to a DRP model, 123
as cause for expansion, 68
as inventory motivator, 26
as modeling constraint, 175
reaction to changes in, 157
decline of, 161
demographic aspects of, 134
expected increase by promotions, 128
forecast in distribution planning, 204
forecasting, 101, 134, 246
as CIMS module, 197
discrepancies, 125
global and local, 124, 181
I.E.'s role in, 135
reliability of, 127
historical data about, 125
in mature markets, 155
monitoring of, 157
price elasticity of, 128
requirements of kanban system, 185
variations, 134, 135
cyclic, 134
seasonal, 134
Deming Prize, 89
Deming, W. Edwards, 35, 89, 193
Depreciation, 65, 103, 104
Design
maintenance considerations in, 185
parallel flows in, 186
redundancy considerations in, 185, 187
stochastic simulation in, 238

Development
as distinct from research, 93
market, 93
process, 93
product, 94
common errors in, 94
oversight function in, 95
Direct costs, 56
Direct labor
as a budget component, 102
decline in importance, 193, 256
maintenance of standards for, 257
standards for, 56
Direct materials standards, 56
Direct resource requirements, 57
Discipline
as budgetary requirement, 100
in CIMS, 195, 218, 230, 242
Japanese, 195
Discounted cash flow return on investment, see DCFROI
Discrete expansions, 74, 75
Distribution, 14, 27, 121, 202
buffer sites for, 124
conflict resolution within, 124, 125
costs in, 14, 121, 181
database for, 14, 126, 130
demand forecasting for, 123, 124
interaction effects, 27, 122
I.E.'s role in, 14, 27, 125
in corporate strategic planning, 14
in integrated planning and control systems, 211
just-in-time approach to, 124, 125
just-in-time sites for, 124
mix of, 126, 181, 202, 204, 207
model for, 181, 203
regional supply sites for, 125
safety stocks in, 124, 125
service considerations in, 121
short-term aspects of, 202
simulation of, 181
statistical analysis of, 127
transportation aspects of, 27, 129
Distribution Requirements Planning, see DRP
Diversification
as a stimulus for growth, 71
as proliferation, 4
in the U.S. steel industry, 3
motivated by market saturation, 133
to reduce cyclical effects, 135
Dome Petroleum, 9
DRP (Distribution Requirements Planning), 123
as CIMS module, 123, 197
data needs, 126
in information flow, 148
priorities in, 124
safety stocks in, 124
DSS, see decision support system
Due dates, 112, 113, 197

Economic order quantity (EOQ), 125, 216
EPA (Environmental Protection Agency), 100
Ergonomics, see human factors
Expansion, 9, 67
anticipation of problems in, 72
as influenced by distribution decisions, 122
as motivated by product positioning, 10
characterization of, 67
discrete, 71, 74
I.E.'s role in, 10
justified by marginal cash flow, 73
marginal, 71
modes of, 71, 75
product mix considerations, 10
proliferation resulting from, 10
successes and failures of, 68
Expected value analysis, 73
Expediting, 113
as a marketing tactic, 250
in CIMS, 199
in poorly scheduled plants, 27, 241
in product development, 168
Experience curve, x, 64
as a measure of capital productivity, 66
as justification for methods research, 89
effect within a single firm, xiii
in budgeting, 98
in product positioning, 138, 157
in strategic planning, xii
monitoring of, 157
theory behind, 138
Exxon, 14

Facilities decisions, 43
capital costs of, 103
improper, 234
in expansion, 5

Facilities decisions (*Continued*)
in strategic planning, 3, 4
in the U.S. steel industry, 3
Facilities utilization, 29, 234
inventory picking in, 30
loading in, 235
maintenace in, 29
materials handling in, 30
plant layout in, 30
Failures of strategic planning
due to excessive structure, 141
due to myopia, 36
lack of follow-through in, 39
management ignorance in, 35
Fatigue of workers, 91
Feasibility study, 69, 70
as part of bottom-up analysis, 37
in acquisition/expansion justification, 69
relationship to capacity analysis, 69
Finance strategy, 12, 97
dividends in, 104
bonds and stocks in, 12, 109
financial theory in, 13, 109
growth as a motivator, 12
I.E.'s role in, 109
in corporate strategic planning, 107
in product pricing, 13
Financial Accounting Standards Board (FASB), 159
Finished goods inventories, 26, 56, 213
Firm orders, 117, 125

Flexibility, 137
as a goal of product research, 87
as a requirement of strategy, 54
as a supply criterion, 117
Flow analysis, 145, 146, 150
Focus of direction, 53
as affected by distribution decisions, 122
C.E.O.'s role in, 5
in corporate strategic planning, 5
in finance, 99
tested against reality, 53
Ford Motor Company, 115
vertical integration in, 13, 111
Forecasting
as a strategic planning tool, x
dynamic aspects, 247
responsibility for, 246
FTC (Federal Trade Commission), 100
Fully allocated costs, 55, 57, 61
in feasibility study, 69
in product positioning, 138

Gantt charts, 228
Goal programming, 205, 207
Goals, 53
Growth considerations, 12, 67, 71

Habit
as a motivator of inventory policies, 214
as a roadblock to acquisition success, 77
as an opponent of success, 44, 147, 154
in organizations, 16
Hallmark Cards, 20
Holding costs, 118, 193
Home computers, example of a young market, 131
Hula hoop, 20, 133
Human factors (Ergonomics), 91

IBM, 11, 83, 131, 132, 136
Implementation of strategic plans
information systems in, 41, 154
line emphasis in, 141, 142, 38
for automation, 184
role of I.E.'s in, 38
short-term aspects of, 22, 154
timetables in, 39
Incompatibility, as a cause of merger failure, 68
Indirect costs, 56, 199
Indirect labor, 102, 148
Indirect resource requirements, 57, 61
Industrial engineers
as coordinators and monitors, 38, 169
as cost control experts, 162
as interlocutors between personnel, xiv, 128, 142
as process-oriented personnel, 4
as productivity experts, 17
at staff level, 18
basic function of, 81
career experience of, xv, 32
current underutilization of, 18, 256
holistic viewpoint of, 5, 32
in acquisition and expansion, 10, 69, 80
in analyzing inventory-related standards, 221

in applying experience curve principles, 139
in bottom-up analysis, 35
in choosing production methods, 25
in computerization of systems, 31
in cost planning and control, 13, 56, 102, 139, 161
in distribution planning, 14, 27, 122, 127, 209
in educating corporate personnel, 43, 103, 161, 245, 248
in facilities utilization, 30
in financial analysis, 108, 110
in flow analysis, 150
in implementation of plans, xiii
in inventory planning and control, 26
in inventory policy, 215
in long-term planning, 41
in lot sizing, 179
in market research, 85
in marketing strategy, 15
in monitoring of costs, xiii
in MRP/CRP systems, 242
in organizational structure questions, 17
in P&C systems, 194
in preventive maintenance, 188
in price elasticity studies, 246
in product development, 21, 95, 163
in product life monitoring, 23
in product mix planning, 24
in scheduling, 27, 223
in selection of scheduling system, 230
in standardization, 96
in strategic planning, 254
in structural analysis, 147
in supply sourcing planning, 14, 111, 114, 115
involvement in R&D, 12, 88
need for initiative from, 257
strategic education of, 19, 55, 257
their role in long-term planning, 18, 21
traditional concerns of, 253
training of, xiii, 19, 26, 27, 30, 75, 92, 93, 180, 191, 233
Infinite-capacity models, 241, 242
Inflation effects on planning, 118, 144, xii
Information flows
analysis of, 150
causes for slowness of, 149
design of, 93
in CIMS, 195, 199
in computerized systems, 31
in distribution planning, 209
in market maturity estimation, 156
in organizational research, 92
Information systems, 195
for demand data, 125
for C.E.O.
for feedback, 247
for P&C systems, 146
for supply analysis, 116
for the study of promotions, 128
I.E.'s role in, 160
in computerization, 199
in cost reduction, 159
in lot sizing, 217
in marketing, 29
in methods of production, 192
in monitoring, 31, 169, 188
in product development, 168
in resolving conflicts, 207
in short-term planning, 21, 154, 159
in structural analysis, 145
large-scale, 159
Integrated planning and control systems, 30, 194
Interaction
as a consequence of expansion, 70
as a consideration in product mix, 173, 179, 236
as a source of unforeseen problems, 75
between scheduling and inventory policies, 219
between strategic objectives, 122
effects of scheduling, 223
in obcuring costs, 71
in two-plant modeling, 181
of CIMS modules, 201
resolution of conflicts from, 206
Intermediate stocks in MRP systems, 118
Internal rate of return, 50, 103
Internal Revenue Service (IRS), 159
Inventories
affected by distribution policy, 121
affected by product positioning, 20
as affected by product mix, 179
as design redundancy, 186
as established by just-in-time procedures, 220
as motivation for planning and control systems, 193
as supply criterion, 114

Inventories (*Continued*)
  counts of, 126, 208, 209
  distributed, 205, 208
  dynamic aspects of, 215
  effect on product mix, 180
  effect on sales, 157
  EOQ approach to, 118
  finished goods, 26, 121
  in master schedule calculations, 177
  Japanese aversion to, 30
  lot sizing in, 214
  minimization of, 30, 119
  minimum level, 177, 220
  monitoring of, 126, 207
  overbooking effect on, 223
  picking, 30
  planning and control of, 213, 25
  raw materials, 26
  safety stock in, 26
  U.S. approach to, 194
  WIP, 26
Inventory control, 193
  affected by distribution decisions, 126
  computerization of, 26
  goals of, 124, 213
  I.E.'s role in, 26
  in distributed systems, 126
  in information flow, 148
  just-in-time variety, 26, 208
  kanban method, 26
  module in CIMS, 26, 197, 198, 199
  reorder point in, 208
  two-bin system in, 208
Investment
  I.E.'s role in, 107
  justification criteria, 50
  life of, 75
  temporizing, 107

Japanese manufacturers
  approach to integrated control, 30
  discipline of, 195
  supply strategy of, 111
Job shop, 112, 194
Juran, J.M., 193
Just-in-time inventory control, 26, 118
Just-in-time production, x, 30, 193, 219

Kanban, 26, 30, 118, 185, 220

Labor intensive operations, 102, 194, 255
Layout of facilities, 234
Le Chatelier's Principle, 154
Lead times
  for supply deliveries, 112
  in distribution decisions, 125
  in master schedule calculations, 177
  in MRP systems, 198
  in RCCP/CRP modeling, 243
Legal staffs, 100
Levitt, Theodore, 4
Life cycle costing (LCC), xiii
Line functions, 88, 148
Linear programming (LP)
  advantages over simulation, 176
  applied to optimizing distribution, 127
  as a strategic planning tool, x
  as a transportation model, 129
  as plant loading models, 176
  in activity analysis, 62
  in product mix analysis, 173
Lite Beer, 75
Lombardi, Vince, 91
Long-term planning
  computerization in, 31
  effect on methods of production, 183
  I.E.'s role in, 41
  planning and control systems in, 194
  vs. strategic planning, 3
Lost orders, 28, 121, 125, 245
Lot sizes, 118, 179, 217
Lot-for-lot ordering, 113, 180
LP, see linear programming

Machiavelli,Nicolo, 77
Machine times
  automated collection of, 183
  standards for, 56, 235
Maintenance
  data needs, 188, 189
  of hardware and software, 224
  strategic policies for, 29
Manager of Market Planning and Development, 93
Managerial accounting, xiii
Manufacturing methods, 88, see methods of production
Manufacturing Resources Planning, see MRP II
Marathon Oil, 9

Marginal cash flow, 186, 190
Marginal expansion, 71
Marginal revenues 106, 191
Market area, 14, 75, 93
Market categories
  mature market, 132
  saturated markets, 132
  young market, 131
Market maturity
  as equipment replacement criterion, 190
  estimation of, 155
  I.E.'s role in, 155
  inventory consequences of, 155
Market research, 68, 75, 166
  as an aspect of R&D, 83, 85
  I.E.'s role in, 85
  in strategic planning, 85
  in-house vs. external, 85
  on demand projection, 101, 136
Marketing
  as CIMS module, 199
  development, 93
  I.E.'s role in, 29, 245
  in corporate strategic planning, 4, 28
  intervention in scheduling by, 222
  primary responsibility for product development, 163
  role in motivating growth, 68
Marketing procedures, 28, 245
  effect on inventories, 28
  information flows in, 29
  product substitution in, 29
Marketing strategy, 15, 131
  activity mix in, 135
  as affected by distribution decisions, 122
  as affected by product life, 133
  based on experience curve, 158
  capacities in, 15
  changes in, 136, 156
  competition's effect on, 15, 135
  customer preferences in, 156
  effect on profits, 28
  for different markets, 15
  I.E.'s role in, 15, 28
  in mature markets, 132
  in response to market changes, 155
  in saturated markets, 133
  operational aspects of, 136
  product costs in, 15
  profitability basis for, 135
  promotions in, 15
  segmentation, 12
  service component of, 135
  short-term changes in, 156
  young market, 131
MARR, see minimum attractive rate of return
Master schedule
  algorithm for, 231
  as CIMS module, 197
  data requiurements, 231
  from DRP demands, 123
  in information flow, 148
  interaction with rough-cut capacity model, 238
  plant loading consequences of, 176
  revision of, 176, 179
  varying interpretations of, 177
  worksheet for, 177
Material Requirements Planning, see MRP and MRP II
Materials handling, 30
Materials research, 84, 87
Mathematical models 63, 166, 206, 210
Mature market planning, 136
Maximum capacities, 45
Measures of effectiveness
  assigned to SBU's, 98
  in organizational research, 92
  in R&D, 84
  short-term, 36
Mergers, 77
  compatability of procedures in, 10
  failures of, 78
  focus in, 9
  I.E.'s role in, 10
Methods of production, 24, 183
  budgeting for, 89
  computers in, 192
  data needed for research in, 89
  design factors in, 183
  I.E.'s role in, 25, 89
  preventive maintenance in, 188
  redundant design in, 187
  replacement/retirement aspects of, 190
  substitution of, 25
Minimum attractive rate of return (MARR), 104
Misplaced orders, 27, 209

Models
calibration of, 62
I.E.'s role in, 181
of aggregate planning, 236
for goal programming, 203
methods in, 62
production planning, 176
Monitoring
in short-term planning implementation, 21
of budget adherence, 102
of CIMS, 200
of corporate functions, 149
of prices, 156
of product development, 166, 168
of production schedules, 146
of quality control, 90
of scheduling systems, 224
MRP, 8, 123
based on unreliable lead times, 113
bill of materials in, 198, 199, 232
data needs for, 232
in cost cutting, 8
in information flow, 148
in inventory scheduling, 220
in production scheduling, 27, 230
in supply scheduling, 118
module in CIMS, 197
need for employee training for, 8
philosophy of, 198
shortcomings of, 23
without CRP, 243
MRP II, 198, 123, 230
Multiple suppliers, 111
Multiple-plant product mix, 180
Murphy's Law, 113
Myopic planning, 36, 52

Net present value (NPV), 50, 75
New charter for Industrial Engineering, 255, 260
Numerical control, 159

Off-budget spending, 92
Oil-and-gas firms, 72, 134
Operating cash, 98, 108
Operations research (OR), 83, 92
Order due dates, 147, 176
Order entry/release
functions of, 145
in sales forecasting, 247
module of CIMS, 199
order tracking procedures, 210
Ordering costs, 118
Organization chart, 141
Organizational structure, 16, 141
accountability aspects of, 17
based on information flows, 144
functional basis, 16, 141
hierarchical basis, 16, 141
in staff sizing, 145
I.E.'s role in, 17, 141, 144
levels of, 141
line activities in, 16
post-merger, 80
proliferation in, 143
research on, 92
Osborne, 131
OSHA (Occupational Safety and Health Agency), 100
Overspecialization of management, xiv

P&C, see planning and control
Paper investments, 79
Parallel design, 187
Pareto's Law, 7, 45, 56, 148, 179, 187
Parkinson's Law, 46
Parkinson, C.N., 143
Penalties in supply contracts, 112
PERT (Project Evaluation and Review Techniques) 164, 165, 168, 233
Planned order receipts in master scheduling, 177
Planning and control (P&C) systems, 193
algorithms for, 219, 248
and inventory control, 213, 215, 218
as information systems, 201
as replacements for manual systems, 195
closed-loop, 198
distribution in, 126
failures of, 195
I.E.'s role in, 200
tailoring of, 198
Plant loading models, 176, 178, 179
Plant within a plant, 75, 76, 137
PM, see preventive maintenance
Politics, 62, 100
Portfolio mix
of conglomerates, 135
of investments, 100
of market segments, 133
speculative aspects of, 191

Preventive maintenance (PM), 188, 189
Price elasticity
  application of, 160
  in promotions, 245
  in short-term product positioning, 160
  monitoring of, 160, 161
  of demand, 128
  time trend of, 128
Price leader in product positioning, 20
Priorities
  by customer groups, 227
  derived from strategic plan, 227
  example of application, 226
  in conflict resolution, 226
  in distribution planning, 205
  in CRP models, 243
  in DRP systems, 124
  in goal programming, 203
  in master scheduling, 231
  in resolving P&C conflicts, 198
  in rough-cut capacity planning, 239
  of Industrial Engineering, 253
  in resolving P&C conflicts, 198
Process concepts, 21, 93
Process development, 93
Process standardization, 95
Process-oriented strategic planning, 4
Product development, 94, 163
  and R&D, 11
  coordination of, 95
  cost estimation, 165
  criteria for, 95
  I.E.'s role in, 21, 94
  in response to market changes, 157
  poor procedures in, 94
  scheduling of, 164, 168
  short-term aspects of, 20, 163, 167
  successes of, 75
Product introduction
  backward explosion of, 171
  coordination and monitoring of, 169
  finalizing special procedures of, 169
  startup manpower for, 139
  studied by aggregate planning models, 237
  timing of, 163
Product life cycle, 21
  effect on market strategy, 133
  effect on short-term plans, 20
  graph of, 22
  I.E.'s role in, 23
  in product development, 163
  in R&D planning, 12
  in short-term planning, 19, 21
  planned, 133
  projection of, 75
  short, 137, 133, 134
Product mix, 23, 172
  and its relation to proliferation, 23
  as a suboptimizing technique, 64
  based on capacity analysis, 46
  bottleneck constraint considerations, 10
  circumvented by Marketing moves, 248
  constraints on, 172
  criteria for, 172
  economic basis for, 58, 174
  I.E.'s role in, 24, 181
  in aggregate planning, 24
  in production planning, 172
  LP formulation of, 175
  market demand considerations, 11
  multiple plant analysis, 180
  optimal, 54, 175
Product phaseout, 23
Product positioning, 155
  based on experience curve, 157
  in reaction to competitors, 156
  market maturity estimation in, 155
  price-cutting tactics in, 19
  short-term aspects of, 19
Product promotion, 161
Product research, 86, 87
Product-oriented strategic planning, 4
Product/process dichotomy, 29, 33
Productivity
  as based on labor hours, 102
  as enhanced by human factors research, 91
  as goal of I.E.'s, 255
  evaluation of, 18
  measures of, 17, 36, 64
  of capital, 103
  of Industrial Engineering, 253, 259
  of sales staff, 250
  white collar, 64
Profit
  as activity mix criterion, 59
  in cash flow, 191
  in ROI calculation, 65
  incremental, 103
  maximization of, 172
Profit centers, 79, 80

Profitability
affected by expediting orders, 251
as a distribution criterion, 122
as a function of downtime, 189
as affected by product mix, 248
as an expansion/acquisition criterion, 72
as the ultimate test of strategy, 54
Profitability analysis
as a prelude to product development, 166
economist's role in, 101
I.E.'s role in, 54, 101
in activity mix selection, 60
in product development, 163
Marketing's role in, 101
price prediction in, 101
Project Evaluation and Review Techniques, see PERT
Project scheduling, 164, 167, 168, 232
Proliferation, 9, 24
as a cause of bottlenecks, 70
as a result of irrational budgeting, 98
as a result of organizational expansion, 16
as a result of poor focus, 7, 53
as legislated against by economic reasoning, 58
of activities, 8
of decision making levels, 143
of function, 141, 149
of hierarchical levels, 141
of products, x
of staff jobs, 144
resulting from incremental decisions, 54
Purchasing module in CIMS, 198
Pure research, 82, 83, 84

Quality control (QC)
activity analysis in, 90
as a category of research, 89
as a reason for Japanese industrial dominance, 89
capacity analysis in, 90
frequency of monitoring in, 90
in American firms, 184
in labor intensive operations, 91
information system for, 117
of supplies, 114
sampling in, 90
workers' response to, 90
Queueing theory explanation of variation effect, 235

R&D, 82
American industry's investment in, 11
as a source of innovation, 83
as a stimulus for market growth, 68
as affected by perception of pure research, 82
costing of, 12
disciplines involved, 12
I.E.'s role in, 12, 84
in corporate strategic planning, 11
internal vs. external, 11
invested in small project groups, 83
management commitment to, 84
materials in, 87
measures of effectiveness in, 84
product research in, 86
responsibility for, 12, 82
role in strategic planning, 82
Rationality of operations, 102, 136, 189, 246
Raw materials inventories, 26, 213
RCCP, see Rough-cut capacity planning
Reactive capability 24, 15 4
Refractory manufacturers, 114
Refrigerator makers, 140
Reject reduction due to quality control, 90
Reliability
as a criterion for kanban operation, 185
as a supply criterion, 112
of materials, 87, 218
of supply lead times, 112
Reliance Electric, 14
Reorder point for inventory, 208
Replacement, 103, 190
cost in investment decisions, 104
cost tradeoff approach to, 191
poor decisions about, 107
rate in saturated markets, 133
Research
as a line function, 88
as part of corporate culture, 88
definition of, 84
in quality control, 90
methods analysis in, 88, 89
on materials, 84
repayment criterion for, 88
Research and Development, see R&D
Research categories, 83, 91
applied research, 84
human factors, 91
materials, 87
methods of production, 89

operations research, 92
organization research, 92
pure research, , 82, 83, 84
quality control, 89, 90
Resource allocation, 61, 179, 258
Resource requirements, 45
demanded by master schedule, 176
derived from models, 176
for marketing, 249
in aggregate planning, 236
in budgeting, 99
in staff sizing, 144
of management systems, 224
prediction of, 61
proration of, 62
Resource standards, 45, 70
Retirement of equipment, 190
Return on investment, see ROI
Revenue estimation, 98, 101
Revenue requirements as investment criterion, 104
Revenues as a limit on expenditures, 100
Rework reduction due to quality control, 90
Risk
acceptance of, 136
as a function of forecast, 135
evaluation of, 73
in diversification, 135
in growth-motivated plans, 67
in investments, 75
minimization of, 135
ROI (Return on Investment)
as a criterion in revenue requirements, 104
as a growth criterion, 68
as a measure of managerial effectiveness, 36
as an investment justification criterion, 50, 73
definitions of, 65
Rough-cut capacity planning (RCCP)
cost of installation, 243
data requirements for, 238, 242
for two interacting products, 240
module in CIMS, 197
numerical example, 239
Rush orders, 113, 117

Safety, 87, 91
Safety stocks
actual, 216
as a function of demand uncertainty, 216
criteria for, 217
for cyclic markets, 137
in distributed inventories, 124
in distribution planning, 208
in inventory policy, 118
in response to marketing criteria, 246
lead times for, 214
lot size effect on, 214
minimization of, 124
of raw materials, 217
policies concerning, 180
Sales promotions, 121, 127, 128, 129
Salesmen's motivations, 28, 222
Saturated markets, 133
SBU's (Strategic Business Units), 55, 98
Scarcity of industrial engineers, 255
Scenarios, 38, 73
Scheduling, 26, 222
affected by distribution policy, 126
affected by facilities decisions, 234, 235, 236
affected by inventory policy, 219
as a supply souce criterion, 112
backward explosion of, 112
Computer Department in, 231
computerization of, 229, 230
criteria for, 177, 232
cost minimization, 226
inventory minimization, 226
maximum flow, 226
tardiness avoidance, 225
data needs in, 27
derived from aggregate planning models, 237
derived from master schedule, 176
disrupted by inventory rules, 180
effect on operations, 230
failures of, 223
I.E.'s role in, 27, 224, 228, 232
in information flows, 148
lead times in, 26
manual, 228
MRP in, 27, 198
need for revisions of, 223
objectives of, 224, 229
of preventive maintenance, 189
of product development, 163
of sequence-dependent jobs, 223
overbooking, 223
priorities in, 228
standardization of, 96
worksheet for, 219
Scrap, 115

SEC (Securities Exchange Commission), 109
Second Law of Thermodynamics, 16
Sequence-dependent jobs scheduling, 223
Setup costs, 118, 194
Short-run products, 137
Short-term debt, 109
Short-term planning, 19, 21, 54, 70, 153, 155, 203
Sideways thinking, 32
Simulation
as a modeling tool, 62
as a scheduling model, 238
as a strategic planning tool, x
as an aggregate planning tool, 237
as an approach to product mix, 176
as a plant loading model, 176
deterministic, 237
in distribution modeling, 122, 127, 203
in selecting inventory policy, 180, 217
stochastic, 238
Skinner, Wickham, 5
Slush funds, 92, 98
Smooth-flow facility, 112, 194
Specialization of management, xiv
Speculation, 137, 190
Spreadsheet software
as a simulation tool, 62
as a strategic planning tool, x
in activity analysis, 62
in distribution planning, 122
in plant loading analysis, 176
in scheduling, 229
Staff functions
capacities of, 61
cost of, 103
growth of, 102
justifications for, 147
lack of line accountability in, 92
standards for, 62
Staff size
for marketing, 249
growth of, 144
I.E.'s role in, 100
planning of, 35, 144
reduction of, 249
Standardization, 95, 96, 172
Standards
and capacity analysis, 45
as a basis for quality control analysis, 116
as a basis for resource planning, 100
as basis for facilities utilization, 235
changes in, 161
development of, 96
for CRP models, 243
for machine requirements, 139
for marketing staff, 250
for staff activities, 148
I.E.'s role in, 149
imitation of in product development, 164
in budget monitoring, 102
in cost analysis, 56
in estimating staff size, 146
in feasibility study, 69
in plant loadin analysis, 176
in retirement/replacement analysis, 191
stabilization of, 162
Statistical analysis, 116, 117, 220
Steel industry
Japanese, 4
U.S.
product strategy in, 4
quality control in, 114
R&D investment in, 82
Steel mills
continuous casters in, 4
Japanese, 4
operation of, 47
special-purpose, 3, 4
supplies for, 114
Stockouts, 180, 245, 246
Strategic Business Units, see SBU's
Strategic planning, ix
criteria for, 122
for Industrial Engineering, 258
goals in, 145
objectives of, 121, 164
Strategic Planning Manager, 143, 144
Structural analysis, 145, 147
Supply categories
order-for-order, 112
routine, 112, 117
rush, 112
Supply sources, 13, 111
acceptance sampling for, 117
as cause of MRP inaccuracy, 198
as cause of quality control problems, 90
economic justification of, 119
flexibility resulting from, 117
I.E.'s role in, 14, 113
in-house, 13, 114

inventory considerations in, 113, 118, 217
Japanese approach to, 13, 111
lead times for, 112
mix of, 120
MRP effect on, 118
multiple, 13, 111
penalties for, 113
premiums for, 113
quality control of, 115
reliability of, 112, 113, 119
review and evaluation of, 120, 218
single suppliers, 13, 111
U.S. approach to, 13, 111
vertical integration in, 13, 111, 114

Tactics, ix, 111
Taylor, Frederick, 90, 194, 255
Texas Instruments, 131
Time value of money, 50, 104
Top-down planning, 34
failures of, 35, 70
Traffic, 145, 199
Transfer pricing, 17, 18
Transportation
analysis of, 130
constraints on, 129
costs of, 129
I.E.'s role in, 129
in distribution planning, 210
modes, 14, 210

U.S. Steel, 9, 46, 147

Variances, 102, 237
detection of causes for, 116, 162
in information flows, 148
in supply quality control, 115
Variation effect, 223
Vertical integration, 13, 111, 114, 119

Washington Post, xi
White collar productivity, 17, 35, 64, 143
Work in process inventories (WIP), 26, 56, 213, 214

Xerox, 75
Zero-Based budgeting (ZBB), 37, 98